# PRACTICAL CHEMISTRY.

Putnam's Elementary Science Series

# PRACTICAL CHEMISTRY

FOR USE IN

SCIENCE CLASSES AND HIGHER AND MIDDLE

CLASS SCHOOLS

BY

J. HOWARD

HEAD MASTER OF THE ISLINGTON SCHOOL OF SCIENCE AND ART

NEW YORK

G. P. PUTNAM'S SONS

FOURTH AVENUE AND TWENTY-THIRD STREET

1873

# CONTENTS.

# PREFACE.

THE present work has been prepared to supply the want —frequently felt by many Teachers and Students of Chemistry, preparing for the examinations of the Science and Art department—of a method, by which the subject could be studied, so as to get a sound knowledge of the elementary facts of the science.

The reports of the eminent men who act as examiners for the Science and Art department have shown, that many students who come up for examination are mere readers, and they rightly decry such a system of *cram*, which can only produce nausea instead of that deep interest in the phenomena surrounding us, which would be produced if science were really studied practically. These views the author fully appreciates. He has for some years taught large classes of young men and boys who have been successful at these examinations, and feels assured that Chemistry cannot be understood unless EXPERIMENT go hand-in-hand with Theory—in fact, that Theory should be built up by the aid of inferences drawn from the experiments themselves.

In addition to this, the Department have offered to assist the individual study of Chemistry by Grants to the Committees towards the expenses of students in a laboratory, and also in extra payments to the teacher. These grants, however, can only be obtained by the student showing at the examinations a good knowledge of Practical Chemistry, and it is believed there is very little chance of getting a *first class* without it.

The teacher, and, if possible, every student, should perform all the experiments described in the book. It will be found to contain all that are suggested in the revised syllabus of the subject for the First or Elementary stage, and many others besides.

The author is indebted to the series of Lectures delivered from time to time by Dr Frankland, and also to his "Lecture Notes on Inorganic Chemistry," for the method of treating the subject, and for many of the explanations of Chemical phenomena given throughout the work. To this work, and to the work on Inorganic Chemistry in the present series, the author would refer the reader for further information on the theoretical part of the subject.

The illustrations in the book are principally from photographs of the actual apparatus used, kindly taken by an old pupil. A few from drawings by another friend.

J. HOWARD.

ISLINGTON SCHOOL OF SCIENCE AND ART,
*October* 1872.

# HINTS TO STUDENTS.

Cleanliness is of the utmost importance. Never put away anything dirty. It takes twice the time to cleanse glass vessels when chemicals have dried on, that it does if washed directly they are done with. Many a young student has thrown away bits of dirty apparatus, which he would have looked upon with pride, had he polished them before putting them away.

Do everything and arrange everything in order. Never be in a hurry or flurry. Serious accidents may occur at times through not keeping cool. Have a place for everything, and put everything in its place. Put all materials and apparatus to be used on your *left* hand before commencing, and move them to your *right* when you have done with them, keeping the middle of your table or bench clear for operating or fitting up. Some arrangement of this kind is absolutely necessary.

Do not be wasteful of your materials, although some of them may be cheap. Pure, and therefore expensive, materials will have to be used in the Advanced Course, and occasionally in the *present*, and the habit of carefulness in all things cannot be too soon acquired. Bad habits, too, are easily contracted, and are difficult to get rid of, hence as few as possible should be acquired.

Look over all preparations required for an experiment before you begin to operate, otherwise you may find the want of something when it is impossible to get it; or you may suddenly discover that you want another hand. It is a good plan, especially for a teacher, to make a list of all apparatus and materials required for the series of experiments which are to engage the attention. Satisfy yourself beforehand WHY you do everything, and never be content with making a thing merely do, if it is not done properly. Badly fitted corks, requiring lutes or sealing-wax to stop the leaks must not be suffered at any time—another should be fitted. Many a serious explosion has occurred for want of these precautions.

In all cases, use the simplest form of apparatus for an experiment. Unnecessary complications are confusing both to teacher and student.

Take careful notes of all experiments as they proceed, on paper, or in a rough note-book. Dr Hofmann used to say, "The scrap of paper well stained with acid is of much greater value than the half-worked-out, though clean notes, written down after the experiment has passed away." The rough notes should be reproduced in a more finished form in a book kept solely for that purpose. The mere copying of scientific facts and formulæ, previously learned in a practical way, is a great help towards remembering them.

Chemical operations should be carried on, if possible, in a room set apart for the purpose. It is better to have it on the ground floor, if possible, as water is more easily laid on, and waste carried off by drains. It must be well ventilated, with a flue in which there is a fire burning, or furnished with a ring of gas-jets, to produce an ascending current. The flue may be furnished with a hood, under which experiments may be performed where noxious vapours arise. A series of zinc tubes, about 2½ inches diameter, arranged over the benches, about a foot from the ceiling, with inverted funnel tubes over the gas lights leading into them, can be arranged to produce an upward current, and will keep the room clear. These may be carried into the flue or out-of-doors.

A chamber about 2 feet by 1½ feet, with glass doors,

should be fitted up in every laboratory, however humble, in connection with the flue, in which experiments with chlorine and sulphuretted hydrogen can be carried on. This, however, can be made to lead out of doors also, by having a Bunsen constantly burning—the waste heat being used for a sand-bath in which evaporations may be carried on, if the chamber be divided by a partition.

The principal sink may be either in the laboratory, or just outside, and should be of glazed stoneware to resist acids. They cost about seven-and-sixpence, and may be had of Doulton, Lambeth. Smaller leaden or japanned iron basins may be let into the benches to carry off waste.

The room must be furnished with working-benches, 2 feet 6 inches broad, round the walls, and, if necessary, down the centre. Each student should be allowed 3 feet 6 inches working space, and a nozzle with a stop-cock should be connected with the gas pipe, for attaching the flexible tube. Narrow shelves should be fixed along the walls, over the benches, to hold the bottles containing *materials* and *reagents*. Cupboards or shelves may be put under the benches, and drawers fitted, to hold the various kinds of apparatus for general use. Each student should keep his own apparatus in a box, which should therefore be large enough to hold all, except such articles that are not liable to be broken if left out. The box may be kept under the bench.

Where a room cannot be set apart for the purpose, flat tables may be laid across the desks in a schoolroom, and bottles, etc., kept in a cabinet or cupboard. A fire should, however, be always burning, so that any unpleasant fumes can be taken to the flue. Water must be at hand in a trough that can be emptied, or in pails.

Of course, where a regular laboratory can be fitted up, further and better means can be adopted than are here suggested; but such arrangements as these will enable a large amount of work to be done with not over-fastidious students. The cost of such an arrangement for fittings, including gas and drainage, would be about £1 per student. The apparatus for general use, such as pneumatic troughs, dessicating bottles, supports, Herepath's blow-pipes, etc., with the bottles for materials, would cost from 5s. to 10s. per student extra—each student being also furnished with such apparatus as recommended in "the Department list."

# PRACTICAL CHEMISTRY.

## CHAPTER I.

MATTER AND FORCE—SIMPLE AND COMPOUND MATTER—NATURE OF FORCES—CHEMICAL FORCE OR AFFINITY—CHEMICAL ACTION—CHEMICAL COMBINATION AND MECHANICAL MIXTURE.

1. THE Earth, with the animals and plants upon it, consists of a variety of substances to which the term **Matter** has been applied. **Matter** is that which has *weight.* Thus it may be proved by a simple experiment with the air-pump, that *air* and other gases have *weight,* and are therefore matter, although invisible. **Matter** consists of minute particles, or *molecules* (little masses, from dim. of L. *moles,* a mass), a term used now to express the smallest portion of a substance that can exist in a separate state. These *molecules* are further supposed by the Chemist to consist of groups of *atoms* (from Gr. *atomos*—*a,* not, and *temno,* to cut), which are indivisible, and which cannot exist in a separate state.

2. **Matter** is either **Simple** or **Compound.** Substances that cannot be separated by any known means into two or more simpler forms of matter, are called **Elements** or *simple substances.* These elements combine together in certain definite proportions, by *weight* and *volume,* to form **Compounds.** There are about **62** known elements, of which about **40** are well known, the rest being chemi-

cal rarities; but there are an infinite number of compounds.

This **Matter,** whether *organic* or *inorganic,* is constantly in motion, either as a whole, or within itself among its *molecules;* being subject to certain **Forces** or *physical agents.* Some of these forces are constantly acting; others are only brought into play under certain conditions; and, lastly, one form of force is frequently convertible into another—as, **mechanical force into heat; heat,** if sufficiently intense, **into light; mechanical force** and **heat** into **Electricity; Electricity** into **Magnetism,** and so on.

Each force has its own special characteristics; but it will only be necessary to mention here those which bear upon the subject of the present treatise. Thus, if a stone be let fall, it comes to the earth, and if not resisted, would reach its centre. The *force* which thus pulls it towards the centre of the earth is an attractive force, called **Gravitation.** It is the attraction of mass for mass, and acts at immense distances, keeping the earth and other planets in their orbits round the Sun, as well as causing water to rush over a precipitous rock and form a roaring cataract, and enabling us to stand firmly upon the ground.

*Experiment* 1. Take a piece of *ice,* its mass is affected by gravity; its particles are held together by an attractive force called **Cohesion.** Pound some of it in a mortar, the cohesive force is loosened by the mechanical force exerted. Put some into a clean *Florence flask,* supported on a *ring-stand* or *tripod,* and apply heat by means of a spirit-lamp or Bunsen's burner. The *solid* ice melts. The cohesive force being further overcome by the *repulsive force* of **Heat**—the molecules being driven farther asunder, and *liquid* water being the result. In a short time the water is converted into steam, which is invisible *in* the flask, but on passing out into the colder air, is converted again into watery vapour—or condenses—as we often see just above the chimney of a locomotive, or a little way from the spout of a tea-kettle.

**3. Cohesion** binds the particles of bodies together, **Heat** drives them asunder. When the cohesive force predominates, we have a **solid;** when the forces are equal, we have a **liquid**; and when **heat** predominates, we have a **vapour, or gas.**

The properties of the ice in the above experiment, however, remain unaltered; it may be made to assume these three states over and over again.

*Exp.* 2. Take a bit of platinum wire about 5 inches long, heat it in the Bunsen's flame, it becomes red-hot; but on cooling, it is found to be unchanged.

*Exp.* 3. Put the wire, or a small fragment, into a test-tube, and pour upon it some nitro-hydrochloric acid (*aqua regia*), made by mixing hydrochloric acid with a few drops of nitric acid, the platinum dissolves, and a pungent smelling gas is evolved. New bodies with totally different properties being produced.

The solution, consisting of platinic chloride, may be put into a small stoppered bottle for future use.

*Exp.* 4. Heat a piece of magnesium wire or ribbon in the same manner as the platinum, it burns with a dazzling white light, and continues to do so when once ignited, giving off a curling wreath of white smoke, and leaving behind a crumbling white substance, called *oxide of magnesium* (the calcined magnesia of the druggist), totally different in its appearance from the metal. If the wire be heated in a porcelain crucible, and care be taken to avoid loss, the resulting compound will be found to weigh more than the metal. The increase in weight being due to the *oxygen* of the air with which it has combined.

The metal is largely used in pyrotechnic displays, and a lamp has been constructed to burn it for photographic purposes.

**4. The Chemical force,** or **affinity,** as it is sometimes **called, has certain** special characters of its own, which distinguish its mode of action from that of other forces. Thus:

***a.* It only acts at inappreciable distances, requiring bodies to be in actual contact.**

*Exp.* 5. Rub a glass rod with a piece of silk, or a stick of sealing-wax on the coat-sleeve. Bring it within a few inches of a suspended lath or a few scraps of blotting paper—attraction takes place. The electric force thus produced is capable of acting at distances that can be measured. Not so chemical force.

*Exp.* 6. Pound some loaf-sugar, and also some potassic chlorate (Chlorate of Potash); mix them together on a plate, and bring a glass rod, moistened with strong sulphuric acid, *near* the mixture—no effect is produced; but the mixture inflames directly the rod is brought into *actual contact* with it. Note the violet flame and the black mass of carbon left on the plate.

To bring bodies into actual contact, it is often necessary to resort to *solution*, *fusion*, or conversion into the state of *vapour* or *gas*, so that a certain amount of mobility may be produced among the molecules. Thus we have two principal methods, the **wet** and the **dry** way. It is not necessary, however, that both substances should be dissolved or fused—one of them being so is often sufficient. A solid should be used in small pieces, or in the state of powder.

*Exp.* 7. Pour vinegar upon some pieces of chalk in a beaker or test-tube, a violent effervescence takes place, and a gas is given off which extinguishes a flaming match.

Ginger-beer and Seidlitz powders, Sherbet, etc., effervescing when water is poured upon them, are instances of mixtures of two different substances, which do not act upon one another until solution is effected.

*Exp.* 8.* Put a few small chips of **Phosphorus** on a plate and cover them with **Iodine,** heat is evolved, and the two substances give off vapours which enter into combination and inflame, violet vapours of ter-iodide of phosphorus being produced.

*b.* **The Chemical force always produces a change of properties**; it is a transforming force, and takes place most energetically between bodies having opposite properties. Numerous instances have already been given, and the student will constantly observe the phenomena.

*c.* **The Chemical force always acts on definite quantities of different kinds of matter.** Thus, when oxygen combines with hydrogen to form water, 16 grammes, ounces, or pounds of the former always combine with 2 grammes, ounces, or pounds of the latter to form 18 grammes, ounces, or pounds. The absolute weight of a body never being altered by this or any other *force*.

Now, it is the province of **Chemistry** to investigate the properties of bodies and the changes they undergo when brought in contact with one another. This involves

* Phosphorus is highly inflammable, burning at all temperatures above the freezing point. It must always be kept in, and *cut* under water.

**Experiment, Observation, Inference**; and hence the term **Practical or Experimental Chemistry.**

**5. Chemical action.** The instances already given of the action of the chemical force, or affinity, show that it exerts itself in various ways; but the *modes of action* will be better understood at a later period.

Some of the differences between mere mechanical mixture and chemical combination will be now alluded to.

Iron, when left exposed to moist air, *rusts*, a reddish-brown powder being the result, which contains both iron and oxygen, the latter being obtained from the air. Zinc, copper, lead, and magnesium also become tarnished; and the metals potassium and sodium become covered immediately with a white crust—in all cases definite compounds being formed, containing a certain weight of metal in combination with a definite weight of the oxygen of the air.

Heat generally assists chemical combination. Thus, if mercury be kept for several hours near its boiling point (360°C), in contact with air, it becomes covered with an orange-coloured film of mercuric oxide, the oxygen of the air combining with the metal under these circumstances, in the proportion of 16 to 200 by weight, the product being the sum of the two, as in all other cases. Without heat, the metal would have been unaffected, whilst the metals platinum and gold, being unaltered by heat, are called noble metals.

*Exp.* 9. Bring a red-hot poker, or iron wire, within a few inches of a small piece of phosphorus, put on a plate, the phosphorus inflames, and white fumes are given off.

*Exp.* 10. Mix together some iron filings and powdered sulphur, this may be done in any proportions—but is only **a mixture, not a chemical compound.**

Draw a magnet through the mixture, or shake some of it over one—the iron filings separate, and the sulphur is left.

Put some of the mixture into a small Cornish crucible, and put it in the fire, the sulphur melts and the iron combines with it in certain proportions. Thus, 56 grammes or grains of iron in this way always combine with 32 grammes or grains of sulphur to form *ferrous sulphide* (sulphide of iron). This sulphide of iron is not attracted by the magnet, and differs in appearance and properties from both the iron and the sulphur.

Gunpowder is a good instance of a *mechanical mixture* —consisting of charcoal, sulphur, and nitre (potassic nitrate). The nitre may be dissolved out by water, and the sulphur afterwards by carbonic disulphide (bisulphide of carbon), the carbon being insoluble. When the powder is fired, an immense volume of white smoke is produced, a variety of new compounds being formed—some solid, others gaseous, and all totally differing in appearance from the gunpowder itself.

Mixtures may be made in any proportions, but "a chemical compound always contains the same elements in the same proportions, however or wherever obtained."

**6. Chemical action generally produces a change of temperature.**

*Exp.* 11. Pour some strong nitric acid on zinc in a test-tube, violent chemical action takes place, the acid is decomposed, giving off ruddy fumes, and the test-tube becomes too hot for the hand to bear.

*Exp.* 12. Put some crystals of sodic sulphate (Glauber's salts) into a beaker, and add some common hydrochloric acid to them. If the bulb of a thermometer be plunged into the mixture, the temperature will be found to suddenly fall. The water locked up in the crystals like ice being converted into a liquid, abstracts heat from everything around it.

**7. Chemical action generally produces a change of state.** Liquids and gases being converted into *solids,* or *vice versâ.*

*Exp.* 13. Dissolve some baric chloride in a test-tube, and some sulphuric acid, a white solid separates, or is precipitated.

*Exp.* 14. Dip a glass rod into hydrochloric acid, and hold it over an open bottle containing a solution of ammonia, white fumes are produced, which if in sufficient quantity would produce the *solid* ammonic chloride (sal ammoniac).

**8. Chemical action often produces a change in colour.**

*Exp.* 15. Add a solution of potassic iodide to a solution of mercuric chloride (corrosive sublimate), both colourless liquids, a splendid scarlet precipitate of *mercuric iodide* will separate.

If potassic iodide be added in excess the precipitate dissolves.

# CHAPTER II.

## WEIGHTS AND MEASURES.

9. MENTION has been made several times in the previous chapter of weight and volume. It will be well therefore before entering on the systematic study of our subject, that the student should know something about the weights and measures used in chemical investigations and calculations.

In England, up to the present time, it has been usual to speak of pounds, ounces, and grains in weight; of yards, feet, and inches in length; of square yards, feet, and inches obtained by squaring the dimensions in linear measure; and of cubic yards, feet, and inches with the gallon, quart, and pint, as measures of capacity.

Thus our *English standard measure of length* is the **yard**, obtained by taking $\frac{1}{1}\frac{2}{3}$ of the length of the pendulum, which beats seconds in the latitude of London (=39·13983). *Our standard of capacity* is the **gallon**, which contains 277,274 cubic inches, being thus related to the cubic foot or yard which contain respectively 1728 and 46,656 cubic inches. The weight of a gallon of distilled water at 62° Fahr. (16·6°C) and 30 inches barometrical pressure divided by 10, gives us the pound avoirdupois, which is our *standard of weight*. That is, a gallon of water weighs 10 lbs., or 70,000 grains. Now, all this is simple enough, but when we come to the multiples and fractional parts of the standards, we have an infinite variety, and the getting up the tables of weights and measures is a most laborious work to the young. Add to this the fact of there being two recognised pounds. The avoirdupois, divided into 16 ounces, and 7000 grains; and the Troy pound, divided into 12 ounces, and 5760 grains, by which we see the lb. av. is heavier than the lb. Troy, whilst the Troy ounce is heavier than the ounce avoirdupois; an ounce of tin being in fact lighter than an ounce of gold, although a pound of the former metal is considerably heavier than a pound of the latter.

In addition to this, the gallon and several other measures and weights differ considerably in different parts of the country. To obtain something like uniformity would hence be a great gain, and if this could be done, much might be said against any radical change.

But the man who would understand the truths of science at the present day must go beyond this. A foreigner makes some important scientific discovery, or clears up some obscure point for the student and teacher who is ever learning. He gives in the course of his statements *weights* and *measures* that it is impossible to translate into their English equivalents without much labour, or recourse to tables which may not be at hand. Hence half the value of the facts is lost for want of an easily convertible system.

For the diffusion of scientific knowledge, then, a universal system of weights and measures would be desirable. It has been proposed to adopt the metric system used by France and many other European nations; and legislation has gone so far already in this country, as to allow of its use side by side with the old system. It is to be taught also in all schools henceforth receiving Government aid, and is now being taught far and wide. Being moreover a decimal system, it is easily learnt.

The foundation of this system is the **Metre,** first obtained by measuring an arc of the meridian—or, taking $\frac{1}{40,000,000}$ of the earth's circumference as the **unit of length.** On subjecting this *arc* to further measurement, the result was found, however, not to be quite the same as that first obtained, but this does not matter, as the **Standard Metre** is now defined to be "the length of a bar of platinum deposited in the archives of Paris," copies of the same being deposited with the warden of the standards in this country.

It stands in relation to our English measure as follows :—

**1 metre** = **39·37079** inches,

or nearly 1 yard 3 inches and 3-eighths.

Taking the metre as the standard of length, the French use the Greek prefixes *deka, hecto, kilo,* and *myria* respectively for the multiples 10, 100, 1000, and 10,000. To mark the fractional parts $\frac{1}{10}$, $\frac{1}{100}$, $\frac{1}{1000}$, the Latin prefixes *deci, centi,* and *milli.* It is therefore a decimal system, and multiplications and divisions are hence performed by "moving the point to the right or left." *See* examples, page 23, first series.

Passing by Square measure as of little interest to the student of chemistry, we come to Cubic measure, or measure of capacity or volume.

Taking a cube with a metre for its side, and taking $\frac{1}{1000}$ of such a cube; or, what is the same thing, taking a cube with a decimetre for its side, we have the **cubic decimetre**, to which the term **Litre** has been applied, and this is the *standard measure of volume or capacity.*

In relation to our English measure, the

| | Cubic inches. | | Pints. | |
|---|---|---|---|---|
| **Litre** = | **61·027051** | = | **1·76** | or about **1¾** pints. |

The litre wine bottle is now quite common in London. The same prefixes are used for multiples and fractional parts, as in the measure of length, thus simplifying the calculations of the system.

Again, taking a cube with $\frac{1}{100}$ of a metre, or a *centimetre* for its side, we have the **cubic centimetre** or millionth part of a cubic metre. The weight of this quantity of distilled water at its greatest density (4° centigrade) is taken as the standard measure of weight, and is called the gramme (Anglicised, *gram*).

In relation to our English weights, the

| | Grains. | Avoir. oz. | Troy oz. | |
|---|---|---|---|---|
| **Gramme** = | **15·43** | **·035** | **·032** | or nearly **15½** grs. |

As in the previous cases, we have decagrams, hectograms, kilograms, and myriagrams; decigrams, centigrams, and milligrams. The gram is about $\frac{1}{1,000,000}$ of an English ton, the kilogram about $2\frac{1}{5}$ lbs. The French use the prefix *demi*, half, in many cases, as *demi-kilogramme*, half a kilogram, a little more than an English pound.

## TABLE OF THE METRIC SYSTEM.

*Standard : The Metre* = 1·0936 *yards*

### LINEAR MEASURE.

| | | | | Metres. |
|---|---|---|---|---|
| Decimetre, dm. | $\frac{1}{10}$ or ·1 of a | Metre. | Decametre, dkm. | 10 |
| Centimetre, cm. | $\frac{1}{100}$ ,, ·01 ,, | | Hectometre, hm. | 100 |
| Millimetre, mm. | $\frac{1}{1000}$ ,, ·001 ,, | | Kilometre, km. | 1,000 |
| | | | Myriametre, mrm. | 10,000 |

Square measure or area is omitted as unimportant to the chemical student.

### CUBIC MEASURE, MEASURE OF CAPACITY OR VOLUME.

*Standard : the Litre or cubic decimetre* = 1·76 pints.

| | | | | |
|---|---|---|---|---|
| Decilitre, dl. | $\frac{1}{10}$ or ·1 of a | Litre. | Decalitre, dkl. | 10 litres |
| Centilitre, cl. | $\frac{1}{100}$ ,, ·01 ,, | | Hectolitre, hl. | 100 ,, |
| Millilitre, ml. | $\frac{1}{1000}$ ,, ·001 ,, | | Kilolitre, kl. | 1,000 ,, |

*Note.*—1 Litre of water = 1 kilogram (2·205 lbs.) A c. c. or millilitre of water 1 gram.

### WEIGHT.

*Standard : the Gramme* = 15·43 grains.

| | | | | |
|---|---|---|---|---|
| Decigram, dgr. | $\frac{1}{10}$ or ·1 of a | Gram. | Decagram, dkgr. | 10 grains |
| Centigram, cgr. | $\frac{1}{100}$ ,, ·01 ,, | | Hectogram, hgr. | 100 ,, |
| Milligram, mgr. | $\frac{1}{1000}$ ,, ·001 ,, | | Kilogram, kgr. | 1,000 ,, |

*The quintal* is 100,000 grammes, and the French *ton* a million grammes.

*Note.*—A grain $\left(\frac{1}{7000}\right.$ of a lb. av. or $\frac{1}{5760}$ lb. Troy$\left.\right)$ = ·0648 of a gram.

1 lb. avoirdupois = ·4536 of a kilogram $\left(\text{about} \frac{2}{5} \text{lbs.}\right)$.

*Examples of Calculations.* N.B.—To multiply by 10, 100, 1000, etc., move the decimal point one, two, or three places to the right, adding ciphers if necessary. Thus 6·75—67·5, 675, 6750.

To divide by 10, 100, 1000, move the decimal point one, two, or three places to the left, prefixing ciphers if necessary. Thus 759·6876 — 75·96876, 7·596876, ·0759687, etc.

SERIES I.

1. How many *metres* in 50 kilometres?
Here 50 × 1000 = 50,000 or 50·, move point 3 places to the right, and add three ciphers = 50000· kilom.
2. How many litres in 62·54 kilolitres?
Here 62·54 × 1000 = 62540· litres. One cipher being required.
3. How many milligrams in 576·596 decagrams?
Here 576·596 × 10,000, or move point 4 places to right, requiring one cipher to be added = 5765960· dekagrams.
4. How many kilolitres in 5765960 litres?
Here, fewer being required, 5765960 ÷ 1000, or move the point which falls on the right of the unit 3 places to the left. *Ans.* 5765·96 kilolitres.
5. How many decagrams in 7·6467 centigrams?
Here 7·6467 ÷ 1000, or move the point 3 places to the left, and as only 1 figure now stands to the left, add 2 ciphers = ·0076467 centigrams.
6. How many cubic decimetres in 54786 decilitres?
Here mark the relation of length cubed to capacity. The cubic decimetre or decimetre cubed, is called a litre. Hence 54786 ÷ 10 = 5478·6 cubic decimetre or litres.
7. Add together 657·6$^{mm}$, 760$^{mm}$, 5·8$^{m}$, 760$^{cm}$, 58$^{dkm}$, 39$^{dcm}$, and subtract from the result, 64·5 metres.
Here, put down all in terms of a metre, thus:—

```
   ·6576
   ·7600
  5·8000
  7·6000
580·0000
  3·9000
--------
598·7176
 64·5000
--------
534·2176
```

As it will be necessary for the student to know how to compare the weights of a given volume of different gases, the plan of Dr Hofmann, adopted by Dr Frankland, will be given here, with some calculations to illustrate it.

The weight of a litre of hydrogen gas at the standard temperature and pressure, *i.e.*, 0°c and 760$^{mm}$ Bar., is

·0896 of a gram. This Dr Hofmann called the crith, and the weight of an equal volume of any other gas at the same temperature and pressure is found by multiplying this by the specific gravity of the gas compared with hydrogen as unity, the answer being given in criths, or absolutely by taking so many times ·0896 gr.

Thus, oxygen being 16, nitrogen 14, chlorine 35·5 ($35\frac{1}{2}$), and air, 14·47 ($14\frac{1}{2}$ nearly), times the weight of hydrogen; the weight of a litre of any of these gases would be so many criths, or ·0896 × 16, 14, 35·5, and 14·47; or, 1·4336, 1·2544, 3·1808, and 1·296512 grams respectively.

SERIES II.

## *Examples.*

*Question* 1. A litre of air weighs (at 0°c and 760$^{mm}$ Bar.) 1·2965 grammes. What is the volume occupied by 1 kilogram?

$$\text{Here } \frac{1000}{1·2965} = 771·3 \text{ litres.}$$

*i.e., Dividing the weight by the weight of a known volume gives the volume.*

2. Water is 773 times the weight of air. A litre or cubic decimetre of water weighs 1000 grams. Find, from these data, the weight of a litre of air.

$$\text{Here } \frac{1000}{773} = 1·2936 \text{ grams.}$$

3. The mercury in a barometer tube is 760$^{mm}$ high (30 inches nearly); the area of the tube is 3 sq. c.m.; and the cistern contains 60 cubic centimetres. What is the cubic contents of the tube and cistern together? Secondly, What weight of mercury would be required (sp. gr. of mercury=13·5)?
   *Answer.* 288 c. c. Weight, 3888 grams=8·609lbs. ch.
4. A cylindrical gas-holder is ·45$^{mm}$ high and 7·0686 $^{sq. dm.}$ area. How many litres of gas will it hold? And how much would the quantity of hydrogen and oxygen gases weigh respectively?
   *Ans.* 31·8087 litres; 2·85 and 45·6 grams.
5. A gas jar is 400$^{mm}$ high and 50$^{mm}$ broad. Required, the cubic contents. (For area × sq. diam. by ·7854.)
   *Ans.* ·7854 of a litre.
6. Another jar is 25$^{cm}$ high and 5$^{cm}$ broad. Required, its capacity in parts of a litre.
   *Ans.* ·490875 litres.
7. Two cylindrical gas jars, each 150$^{mm}$ high and 40$^{mm}$ broad. What volume of gas will the two hold?
   *Ans.* ·376992 of a litre.

8. How many criths in 50 litres of hydrogen, oxygen, nitrogen, air, and chlorine respectively? (*See* above data.)
*Ans.* 50, 800, 750, 725, 1775.

It may be useful to add a few facts by way of comparing our present system of weights and measures with the *metric system*, with some examples showing the method of conversion.

The cheap boxes of weights usually sold by instrument-makers usually contain *grains*, *scruples*, and *drachms*, apothecaries' weight, which is equivalent to the Troy weight. Cheap boxes of *gram* weights sufficiently exact for elementary purposes are, however, now to be had.* Liquid measures are in almost everybody's possession. These are usually graduated into fluid ounces. An imperial pint contains 20 fluid ounces, a fluid ounce of distilled water weighing exactly 1 ounce avoirdupois, or 437½ grains. Some measures are graduated into cubic inches. A cubic inch of water weighs 252·458 grains (nearly 252½). A *litre* wine bottle is easily obtained for comparison with our liquid measure.

The imperial pint = ·5675 of a litre, or 34·66 cub. in.

The imperial gallon = 4·54 litres.

This quantity of distilled water at 62° F. weighs 10 lbs. av.

The cubic foot = 28·315 litres or cubic decimetres.

SERIES III.

*Examples to be worked out or verified by Experiment.*

1. Weigh out 10 grains of sand. Find the number of grams.
2. Weigh out 1 scruple of sand. Find the number of grams.
3. Weigh out 20 grams of zinc, and find how many grains.
4. Measure out 10 fluid oz. of water, and find what part of a litre it is.
5. Measure out 5 fluid oz. of water. How much would it weigh at 62° F.?
6. Measure out 1 litre of water. What does it weigh in grams and grains.

---

* A shilling box of such weights can be obtained at Maw's, in Aldersgate Street, London, from 5 milligrams to 10 grams. The gram weight is only 2 milligrams out.

7. What is the weight of a pint of water in grams, and its volume in parts of a litre?
8. What is the weight of a cubic inch of water in grams?
9. How many cubic inches in 3 pints of water? What is the volume in terms of the metric system?
10. A gas-holder contains 6 gallons of water. How many cubic feet, and how many litres of gas will it hold?
    *Ans.* ·96275 cubic feet, or 27·26064 litres.
11. What is the weight, in grams, of 2 scruples of potassic chlorate?
12. What is the weight, in grams and kilograms, of 580 litres of water?

---

## CHAPTER III.

### Apparatus and Preliminary Operations.

**10.** Before commencing the systematic study of the subject, the student should make himself acquainted with the apparatus to be used, and the modes of fitting up certain parts. This can be better done by making and fitting up as many as possible himself, than by his buying them ready made and fitted. He should not, however, spend valuable time in making and trying to mend apparatus that can be easily obtained at the instrument-maker's for a small sum.

**11. Glass-working.**—Much of the apparatus used by the chemist is made of glass, which, on account of its plastic nature when hot, enables it to take any form impressed upon it at that time. Glass tubes bent into various shapes are in constant requisition, and the student should acquire dexterity in making these for himself. Some of the glass tubing imported from Germany is very difficult to fuse.* It is hence called *hard* glass. This is

* "Difficultly fusible," as it is termed, not very euphoniously—and the same of "difficultly soluble." These terms were, however, used by Faraday and Miller.

only useful for combustion-tubes, which are required to stand great heat. There is another kind which is more easily worked; it is readily fused or brought into the plastic condition, and hence called *soft* glass—this should be selected. English tubing is soft, but it contains lead, which is easily *reduced* on heating. French glass is also soft, but the German soft glass is the best for fitting up apparatus. The tube should be of moderate thickness; if too thin, it bends suddenly when heated; but if too thick, it is difficult to work. It is to be had of various sizes. These are the most useful.

The best flame for bending ordinary tubes is a fish-tail gas burner, but the flame of a spirit-lamp will do, and a small tube may even be bent in a candle flame. A fish-tail burner fitted on a telescope tube with a cast-iron foot, so that it may be raised to various heights to suit different persons, is a valuable adjunct to a laboratory for bending tubes, as the ordinary gas lights are often too high. If gas be used, the soot deposited may sometimes stain the glass, if heated too long, but does no injury, and can generally be wiped off. Heat about an inch of the tube where it is wished to be bent, holding it with both hands, and turning it constantly so as to heat it regularly. When it is found by the pressure of the hands to be soft and yielding, remove it from the flame, and gently bend it with the concave part from you, taking care to *keep it in the same place.* If the *concave* portion be made too hot, it is apt to buckle; if, on the other hand, the *convex* part be too much heated, the tube is apt to become flattened and contracted in its channel in the course of bending. Of the two, it is better that the convex portion be made hotter than the other. When finished, tubes should be annealed by gradually withdrawing them from the flame, not cooling them suddenly. Never put a hot tube down on the bench, but support it on some badly conducting material till cold.

If large tubes are to be bent, a **Herepath's** blow-pipe may be used with advantage, the glass being brought gradually into the outer flame from beyond; or a charcoal fire may be substituted. Glass begins to soften before a visible red heat, but the manipulator depends more upon the touch than the eye.

Small tubes about three or four inches long, for heating substances over the **Bunsen**, or blow-pipe flame, are easily made from a bit of hard combustion tubing by means of the Herepath, in the following manner:—

Take a piece of tubing about $\frac{3}{16}$ of an inch in diameter, and double the length of the required tube; heat it in the centre, gradually at first, then strongly, and when quite soft, pull it out suddenly, and you have two tubes. The little adherent piece is easily removed by the scratch of a file, and the point fused afresh.

The following extracts are from the late Professor Faraday's admirable work on Chemical Manipulation.

**12. Blowing Bulbs.**—"A more delicate operation than any yet described, and one that requires considerable practice for its performance with even moderate success, is the blowing of bulbs, and other expansions, either in the middle or at the end of a tube.

"Facility will be best obtained by practising with a piece of tube about nine or ten inches in length, one-third of an inch in diameter externally, and one-tenth of an inch in internal diameter. The end is first to be closed, and then about two-thirds of an inch in length of the closed extremity is to be uniformly heated until so soft as to bend from side to side by its own weight. The aperture is immediately to be placed between the lips, and by means of the mouth, air is to be propelled for the purpose of expanding the soft glass. This must be done quickly, but cautiously; and as soon as the eye, which must be constantly fixed upon the heated end, perceives enlargement there, the force exerted by the mouth should be slightly diminished, and the operator should hold himself ready for its instantaneous cessation. For if the power which expands the glass at first be continued in full force, the glass will suddenly burst out into a large irregular thin bubble of no use (*fly to pieces, and form the so-called 'frost'*). This follows as a natural consequence, for every enlargement of the bubble diminishes its thickness, and consequently, its resistance, and at the same time increases its internal area, and in that respect increases the power of the air impelled into it; and if the enlargement take place quickly, so that the glass has not time to cool, it cannot but happen that the bubble should expand to a large size. To avoid this, air should be thrown in from the mouth only, and not from the lungs: as the glass expands, the force with which the air is impelled should be diminished; and the operator should not endeavour to finish the bulb at once, but having succeeded in expanding it to such a size that the internal diameter is five or six times the thickness of the glass, he should heat it again, and complete the bulb at a second operation.

"The glass must never be blown whilst in the flame, but is always to be expanded in the air; as it swells, it must be turned, and the thinnest parts brought to the lowest position.

"If an expansion be required in the middle, the tube, if not closed at one end, must be temporarily closed with the finger or a cork. The glass is then to be heated where the bulb is required, and then blown out as before."

The cut ends of tubes should always be fused, so that they shall not cut the fingers or india-rubber connectors. Heat gradually, about half-an-inch from the end, and then the end itself, till the edges begin to glow.

To seal a platinum wire into a glass tube requires some practice, and is not much wanted by a student at first. It can be done with platinum, because the rates of expansion and contraction of that metal are the same as that of glass.

**13. Glass-cutting.**—Glass tubes are easily cut, by making a notch with a *sharp, three-cornered file*, turning the glass rather than moving the file. Break, by grasping with both hands, and bending with the notch away from you, as you would snap a dry stick. If the tube be large, carry the file-mark round, and make the notch deeper on the side farthest from you.

Sometimes the neck of a flask we wish to keep gets broken. The point of a hot iron passed round the tube will frequently lead the crack, and the rough edges may then be fused, as mentioned before. A flask, or retort neck, may sometimes be cracked round, by tying a string previously soaked in turpentine round the place, and then setting fire to it, turning the flask the while. When burnt out, invert the flask and dip it into water up to the point; it will generally crack at the place desired. Glass may be cut with a diamond; but it is not likely the ordinary student will possess one. A scratch, however, is of no use with this substance; it must be cut.

**14. Lamps.**—Where gas is to be procured, the **Bunsen's Burner** is the best source of heat for chemical operations, but spirit lamps may be substituted where gas is not laid on.

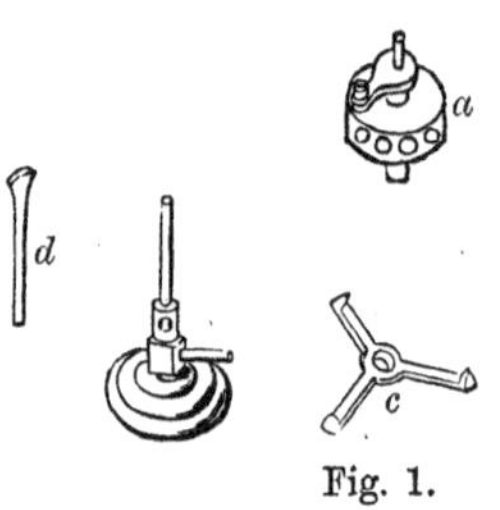

Fig. 1.

The Bunsen's burner is supplied with a heavy cast-iron foot, into which is screwed a small piece of tube to connect it with the supply-pipe by means of about 2 feet of the brown or vulcanised india-rubber tubing $\frac{3}{8}$-inch diameter. The burner is screwed to the top of the foot, and a piece of brass tube is screwed on over this, perforated with a somewhat large hole on each side of the base, and supplied with a moveable ring to close the holes, if necessary. The air entering the holes is mixed with the gas as it issues from the jet, and complete combustion takes place at the top of the tube, a very hot, non-luminous, smokeless flame being the result. If the pressure of gas is too small, the air mixes with the gas in the tube and explodes it near the jet, without danger. If, however, the pressure is too great, and it is necessary to reduce it, the sliding ring may be so moved as to regulate the supply of air.

If the brass tube of the *Bunsen* have a worm at the top, it may be fitted with a brass cylinder, having a piece of fine wire-gauze at the top, the base being perforated with holes for the admission of air. This gives a large, hot, smokeless flame for flasks, and is called a **Gauze Burner.**

Fig. 2.

The **Rose Burner** (*a*, Fig. 1), to screw on to the top of the Bunsen, is another arrangement for the same purpose.

The *Chimney* (*b*), with *star support* (*c*), is sometimes useful.

**15. The Blow-pipe Jet** (*d*) is a piece of brass tube, flattened at the top, which drops into the tube of the Bunsen and gives a thin, luminous flame, to be used with that instrument.

A spirit-lamp, with circular wick, or a Berzelius' lamp, in which the spirit reservoir is at a distance from the flame, and made to slide up and down the rod of a ring-stand, should be furnished to every laboratory where gas cannot be had.

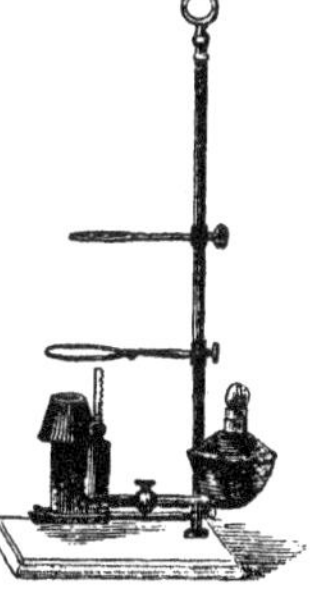

Fig. 3.

**16. The Blow-pipe.**—Black's form is the best, having a reservoir for the air at the bottom, which also furnishes a means for retaining the moisture (*a*). Another form, made of brass, is very useful for retaining the moisture in a sphere, midway between the jet and the mouth. It is shown in Fig. 4, (*b*). The sphere is made in two pieces. It may be furnished with a bone mouth-piece.

*To use the Blow-pipe.*—The student must *not* get the notion that he is to empty his lungs of air and then gasp for breath. He must, in the first place, inspire and expire through the nose. Placing the blow-pipe between his lips, and filling his mouth with air from the lungs, he must learn to use the muscles of the mouth to regulate and sometimes to supply air to the instrument. The mouth, in fact, acting the part of a distended bag, into which the air is taken by mouthfuls, the muscles of the cheeks and jaws propelling it forward so as to produce a constant blast. It is not necessary to produce a strong current, the smaller stream being as efficient as a larger one.

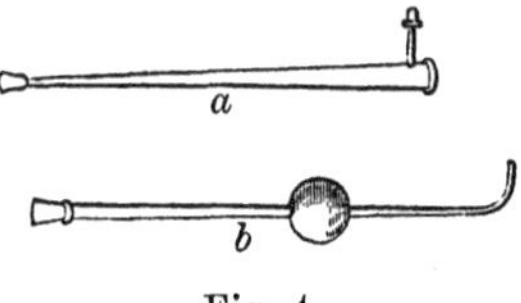

Fig. 4.

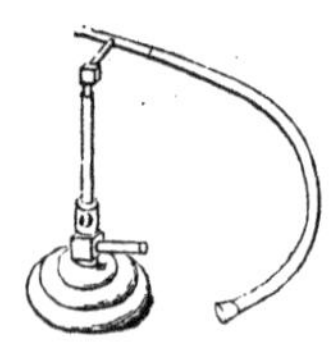
Fig. 5.

**17. Herepath's Gas Blow-pipe.**—The jet, furnished with flexible tube and mouth-piece for sending the blast of air, is usually fitted to a Bunsen; gas being supplied by the latter.

**18. Supports.**—These should not be selected for show. They are requisite to hold apparatus in various positions, and substitutes may mostly be made or fitted up for the occasion.

**19. The Retort Stand,** having a heavy foot, and furnished with 3 rings, and clips of various sizes for test-tubes when they have to be heated for some time. Some of the larger ones are furnished with a screw clamp, lined with cork, to hold a flask or tube containing mercury. A very good substitute is made with an iron rod fixed into a wooden base, with some pieces of thick wire for rings.

**20. The Tripod** may sometimes be used as a substitute

for the above. Both are usually furnished with a piece of wire gauze to lay on the top; a triangle to hold an evaporating dish, when it may be safely exposed to the direct flame; and a sand bath.

**21. The Lamp Furnace** consists of an earthenware cylinder to cover a Bunsen and form a chamber of hot air, and is a useful addition. It is usually furnished with a dish at the top, to be used as a water-bath for heating substances that do not require a temperature above the boiling point of water.

**22. The Table Support** consists of a flat table fixed on a rod which slides up and down a tube rising from a wooden foot. The tube is furnished with a screw, so that the table can be fixed at any height.

The wooden crook (*a*), and one with 3 pegs (*b*), fixed into a ball, are useful, and may be fitted to the same stand.

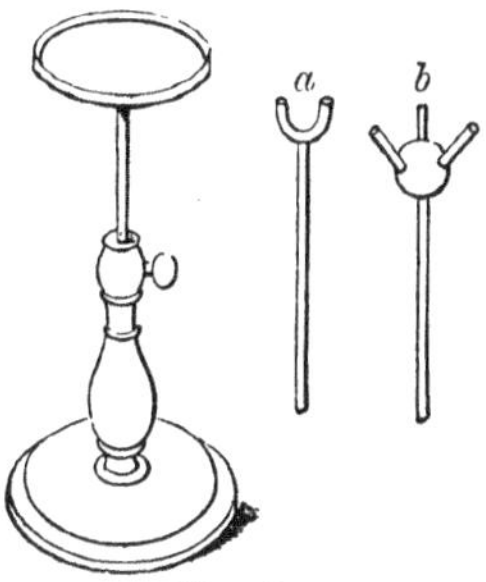

Fig. 6.

A few wooden blocks, 3 or 4 inches square, and from ½ an inch to 3 inches high, are handy, and would in many instances do quite as well as the more finished support.

**23. The Wooden Vice**, lined with cork (Fig. 7), is an almost indispensable requisite.

**24. The Pneumatic Trough**, for the collection and transfer of gases over water, is made either of japanned tin plate or earthenware. The larger kinds, as in Fig. 8, have 2 shelves for supporting gas jars. The small ones have a shelf for supporting the gas jars, perforated with two holes, having inverted funnels below them to lead the gas into the jars. (*See* Fig. 8-*a*, page 34.)

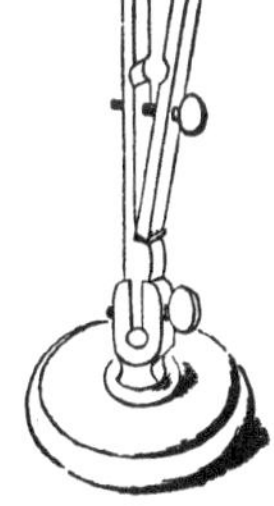

Fig. 7.

The stoneware trough, Fig. 9, is furnished with a beehive shelf, for the same purpose as the funnels in that first named. Glass dishes, for smaller quantities of water, are useful on the lecture table, on account of their transparency; but the student in the laboratory may substitute basins instead.

A substitute for a pneumatic trough may be made as

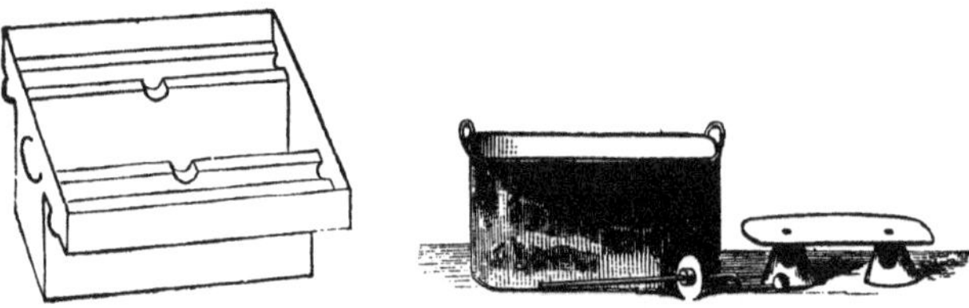

Fig. 8. Fig. 8-*a*.

follows. Take a flower-pot, make a hole in its side, and place it upside down in a wash-hand basin. Invert the gas jar (filled with water), over the hole in the bottom of the pot, and bring the delivery - tube through the hole in the side.

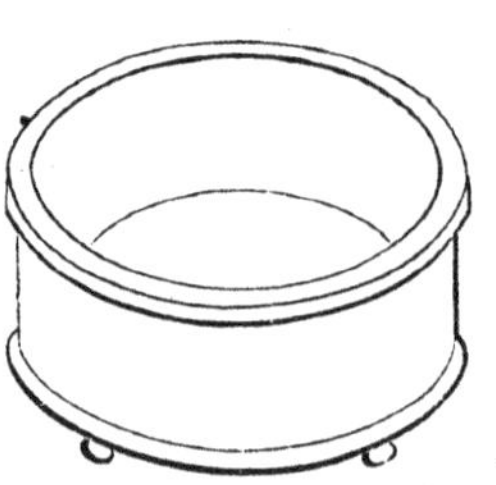

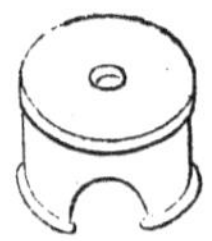

Fig. 9.

The gas will rise through the hole in the bottom into the jar. The student should practise filling jars with water, inverting them in the trough, and then filling them by means of a glass tube with *air* from his *lungs*. He may also learn to transfer air from one jar to another filled with water, by depressing the jar of air with his right hand till its mouth is just below the edge of the other, then, gently tilting it, and allowing bubbles of air to rise into the other jar. Do this over and over again, till skill is acquired. He may also fill a *cylindrical jar* with *coal gas* by means of a flexible tube, with a piece of glass tube attached to dip just under the surface of the water. Grease a glass plate, bring it below the surface of the jar, slide it over the mouth, raising it out of the water and standing it upright on the table. Take a lighted taper in the right hand, and slide off the greased plate with the left, apply the light, and the gas will inflame without explosion, if care has been taken not to mix air with it. Air may be mixed with gas in another jar, and the flame will cause the gas to explode without injury to the *cylinder*, as no obstacle is presented to the expanded gases rushing out. Collect a jar of coal gas, without admixture with air, and inflame it while held in the left hand in an inverted position. (*See* Fig. 23, page 69.) The gas does not escape, because it is lighter than air. The student may also transfer a jar of coal gas, collected at the pneumatic trough, upwards into an empty jar, held as in Fig. 24, page 69, the gas rising through the air and displacing it on account of its lightness. Test carefully with a light. A slight explosion will not injure *a cylindrical jar* of moderate thickness.

Practice of this kind will be found very useful, and may be much extended by the teacher and thoughtful student, and manipulating power will be gained of great value in future experiments.

The gas-holder will be described in the chapter on oxygen.

24-*a*. **The Mercury Trough** will perhaps sometimes be used by the teacher at his lecture; but, on account of the expense of the material, and its liability to loss, one as small as possible may be selected. One to work with 4 lbs. of mercury is to be had of Jackson, Barbican. It requires careful practice to get used to filling tubes and small jars at the mercury trough, and should first be done with no one fidgeting about. *Globules* spilt may be run into *one* and be swept down by the operator, so that little loss need be really sustained. The mercury trough should, however, *always* stand in a tray or other vessel when used.

**25. A Wedgewood Pestle and Mortar** is required for pulverising substances so that they may be more freely attacked by chemical reagents. Very hard substances, like *sulphide of iron*, should first be reduced to fragments by wrapping them in paper and striking them a few blows with a hammer on a stone or anvil.

A selection of sound corks will be necessary. Before using them, they should be pressed, or put in paper and rolled under the foot to soften them.

**26. Cork-borers** consist of 3 or 4 brass tubes, sharpened at one end, and having a rim at the other, with a hole through the upper part to admit of a stout wire being put through to give greater purchase. They may be sharpened when necessary with a fine file. *A rat's tail rasp and file*, the first for enlarging a hole made by the cork-borer, and the latter for making it smooth, should also be possessed by each student.

---

## CHAPTER IV.

### Operations Continued.

There are certain operations of constant occurrence in the *Advanced course* which it will be well for the beginner to know something about at once, as they occur occasionally in the *First stage*, when more important matters are pressing for attention than the mere operation itself.

**27. Solution.**—This follows naturally after pulverisation, mentioned at the close of last chapter. It gives greater mobility to the particles of bodies, preparing them for, or producing *chemical action*, sometimes separating a *soluble* substance from one that is *insoluble*. If by dissolving a substance, no alteration of properties is produced, and on evaporation the solid is again obtained, it is a *Simple solution*. If, on the other hand, *chemical action* takes place, the student must know beforehand the nature of it, as new compounds will be formed. In a simple solution, the body exists in a *free state*.

**28. Water** is the best *Solvent*, and is always tried first. Most substances are more soluble in *hot* than in *cold* water, so that if a certain quantity dissolve in cold, more will dissolve if the water be heated, the substance, however, frequently separating in a crystalline state as the liquid cools.

**29. Salt** is equally soluble in hot and cold water, but *lime* and *magnesia* are somewhat *less soluble* in hot than in cold.

**30. Acids** are usually tried next to water; hydrochloric being preferred because of the solubility of most of its compounds; then nitric, which has a great tendency to part with oxygen; and lastly, but rarely, sulphuric.

**31. Alcohol** comes next as a solvent, and solutions in this liquid are called tinctures. *Methylated Spirit* will do in most cases.

**32.** If a substance is insoluble, it is tasteless, but it is not always safe to taste an unknown substance. Loaf sugar and sugar-of-lead (Plumbic acetate), are similar in appearance, though not in weight, but the latter is *poisonous*. If a substance is soluble, and a bit be suspended in water by a thread, streams of the dissolved substance will be seen* to descend on account of its greater density.

**Solution** is generally performed in *test-tubes*, *flasks*, or *beakers*. *Flasks* are used when much vapour is evolved,

* Owing to its different refractive power. In this way the hot air rising above a burning gas jet is visible passing through the denser air.

and to drive off the greater part of the water when a substance has to be evaporated afterwards to dryness. They should not be put on a table when hot; they may be supported on wire rings covered with *list*, or, in one of the straw rings from the bottom of the covering of a Florence oil flask.

*Beakers* are thin cylinders of glass with flat bottoms, and are sold in *nests* of 3, 4, or 6 of different sizes. Being thin, they bear sudden changes of temperature well, and from their width admit of the stirring-rod being used *with caution.* They may be supported on wire-gauze, or a sand-bath on a tripod. Be careful to wipe off the moisture that first condenses on the outside of such thin vessels, or they will be sure to crack. *Test-tubes* may be held when hot with a band of folded paper passed round them.

A cold liquid should not be added to another when boiling, or, if absolutely necessary, only in small quantities at a time, either through a funnel or down a glass rod previously wetted. On no account must it be poured down the side of the glass.

If the whole of a liquid is to be driven off, the larger portion may be evaporated in a flask or beaker, and the remainder can then be poured into an *Evaporating dish.*

**33. Evaporating Dishes** should not be exposed to the naked flame, or at all events, the flame must not be allowed to pass *above* the line of the liquid, or fracture will result. They may be supported on wire gauze or a sand-bath. Towards the end of an evaporation, the substance becomes pasty and spits. It should then be watched, and before the last drop or two has evaporated, the lamp should be taken away, the heat of the vessel being sufficient to complete it.

The *water-bath* may be used for evaporations where the substance would be injured by a temperature above the boiling point.

Evaporation is often the first step to crystallisation.

**34. Crystallisation** often takes place when a substance passes from the fluid to the solid state, taking up regular forms. Thus *common salt* always crystallises in *cubes*, chlorate of potash in *flat tables;* sulphate of zinc in *needles*. To facilitate the process, the liquid must be allowed to evaporate slowly. Crystals frequently separate from a hot saturated solution as the liquid cools. Very slight changes in the physical condition of a liquid cause crystallisation to begin. A *nucleus*, or something to attach itself to, seems often to start it. Thus if a hot saturated solution of sodic sulphate (Glauber's salts) be allowed to remain *at rest* in a small flask, crystallisation does not take place, but the least motion, the touching of the surface with a glass rod, or the dropping in of a crystal, is sufficient to start it, and *crystals* dart through the liquid which becomes a solid mass, the water being retained in the state of ice (water of crystallisation).

*Exp.* 1. Dissolve some common alum in a large boiling tube (large test-tube), heating and adding the substance till no more will dissolve. Stretch some strings or bits of worsted across a saucer or other shallow vessel, pour the solution of alum into it, and put it aside in a warm place. Or, a bit of coke may be put into the solution. Crystals like double pyramids will form around it.

Large crystals may be obtained by selecting some of the most perfect and putting them into a separate vessel containing a saturated solution of the salt, turning them over daily with a piece of stick on to a different face, replacing the liquid every few days by a fresh cold saturated solution, after draining off the "mother liquid." Substances which crystallise in the same forms, and are similar in atomic constitution, are said to be *isomorphous*. Thus chrome alum, though not the same in composition as common alum, crystallises in the same form, and very pretty crystals may be obtained by adding a solution of this substance to that mentioned in the above experiment.

Some substances crystallise in two forms, and are hence *dimorphous*. Thus sulphur crystallises from the state of fusion in needles, but from its solution in bisulphide of carbon in octahedrons.

Substances which do not assume the crystalline state are said to be *amorphous*.

**35. Filters and Filtration.**—Packets of filter paper may

be purchased of the proper sizes to fit 2″ and 3″ funnels (the most useful sizes), which are used to support them. Stout white blotting-paper may be used, or the filter paper may be bought in sheets and cut to sizes by means of circular discs of tin plate or Mohr's filter cutters. To fix the filter, fold the filter paper in half, then again at right angles, and open out by leaving *three* thicknesses on one side and one on the other, as in *a*, *b*, *c*.

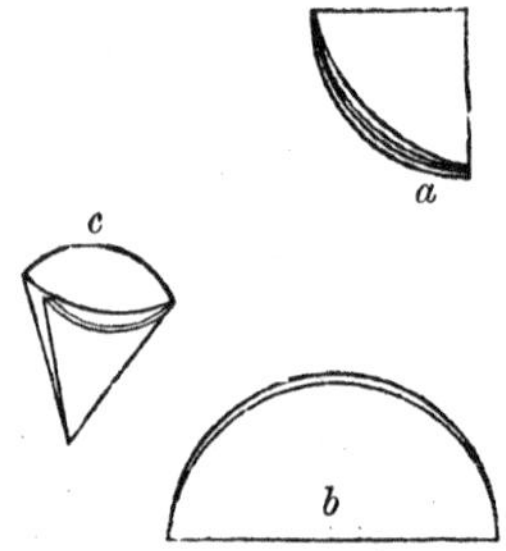


The filter is to be put into a glass funnel, which should be made with its sides sloping at an angle of 60°, in order that circular filters may fit accurately. The funnel may be supported on the ring of a retort stand, or by a triangle on a tripod over a beaker. The end of the funnel tube should enter or touch the side of the beaker to prevent splashing.

**36. Decantation** or pouring off may sometimes be resorted to instead of filtration.

*Exp.* 2. Mix together some sand and salt. Fold a piece of writing paper, and shoot some of the mixture into a dry test-tube. Half fill with water and shake, the salt will dissolve. Fold a filter as above directed and put it in a glass funnel, supported by a ring of the retort stand with a beaker below. Wet the side of the filter by a stream of water projected from the wash-bottle afterwards described. Pour the whole of the contents of the test-tube carefully on the filter; the clear liquid called the *filtrate* which runs through will contain the salt in solution, which may be tested by its taste.

Liquids may be poured down a wetted rod into a funnel in the way previously described. The student may practise the separation of the constituents of gunpowder as mentioned in Chap. I.; or, at all events, he may dissolve out the nitre and test it.

**37. The Wash-Bottle** consists of a 30-oz. German flask, or a Winchester pint will answer very well if it can be got with a wide mouth. A sound cork must be fitted with two tubes bent in the following manner. One tube must be bent at an obtuse angle, and be just passed through the cork. The other must pass through the cork to the bottom of the flask, and must be bent externally at an acute angle, and terminate in a jet as in Fig. 10.

The bottle being filled with water, and the mouth applied to the first tube, air is to be forced in, and the increased pressure drives the water up the second tube and out at the jet. Its use is to wet a filter and to wash down particles from its sides. The jet may be attached by a piece of flexible tubing, so that it can be turned about at pleasure, without shifting the position of the bottle.

Fig. 10.

**38. Distillation** is not much required in Inorganic Chemistry. It consists of separating one *fluid* from another, the former being more volatile; or in separating a fluid from a solid, the former being wanted and the solid not.

It is carried on best in *retorts.* These should be light and free from specks or flaws, which are liable to make them crack when heat is applied. If stoppered, the opening called the tubulature should be in the centre of the body, so that substances poured in shall not soil the neck. It frequently happens that more materials have to be added whilst the operation is proceeding. If so, the powder should be added gradually, being shot in from a piece of *glazed paper*, and if liquid, by means of a thistle-headed funnel. It is better, however, to stop the operation to add new materials, and shake or whirl the retort round, as solid substances are apt to sink to the bottom. The tube of the retort is generally passed into a *globular receiver* made for the purpose, but a clean flask of any kind will do, so that the neck is wide enough to admit the tube of the retort. No cork is needed. Cold water must either be allowed to drip on the *receiver* when distillation is going on, or wet cloths or blotting-paper must be put on it to condense the liquid, called the *distillate*, which comes over.

*Exp.* 3. Put half a gill of port wine into a 2-oz. retort, and pass the tube into a *receiver* or Florence flask standing in a basin containing water, covering the latter with a wet cloth or some wetted blotting-paper. Heat the retort on a sand-bath, vapours will pass over and condense in the receiver, which by their spirituous taste will be recognised as *brandy*. The residue in the flask is little more than coloured water. If the distillation be carried on too long, water will evaporate also, but the spirit being more volatile, comes over first.

Fig. 11.

*Cognac Brandy* is distilled from the *lees* or *refuse* in the wine manufacture.

**39. Clark's Retort and Receiver** is handy for distilling small quantities, and will be mentioned under the head of nitric acid. (Fig. 40, page 119.)

**40.** For distilling large quantities the chemist uses a *Liebig's condenser*, where the condensing surface is increased by surrounding a long tube by a wide tube containing water, called a water-jacket.

**41. Sublimation** differs from distillation in the nature of the product, being a solid, as in the refining of crude sulphur. The materials are melted, and the sulphur being volatile, is allowed to pass into a cool chamber, where it condenses in the form of *flowers of sulphur*.

*Exp.* 4. Put a few grains of *iodine* into a clean Florence flask and heat gently; the substance volatilises, violet vapours filling the flask and cooling on the sides in a crystalline form.

# CHAPTER V.

## ACTION OF THE ALKALI METALS POTASSIUM AND SODIUM ON WATER, HYDROCHLORIC ACID, AND AMMONIA.

**42. Water** was called by the ancients one of the *elements*, although, doubtless, they did not use the term with the same meaning as we do now. Let us endeavour to understand something further about its composition.

On page 14, we learned that it could be converted by heat into invisible steam: this steam occupies about 1700 times the space of the water that produced it—but its molecules, though repelled from each other by the force of heat, are still water. The intense heat, however, of a succession of electric sparks passed through steam separates its molecules into gases, in other words, decomposes the steam. The experiment, however, is somewhat too difficult for a beginner to perform, and requires an *induction coil.*

Some metals have no effect upon water at any temperature, as gold, silver, and platinum; others, as copper, iron, and tin, decompose it at a *red heat;* whilst the metals potassium and sodium decompose it at ordinary temperatures.

*Exp.* 1. Get some of the metal *potassium*, which is prepared from wood ashes. It must be kept in petroleum, as this liquid contains no oxygen, for which it has a great affinity. It is *soft*, and when cut with a knife, is bluish-white. Take a bit the size of a pea, throw it into a basin or little glass dish containing water; it will ignite, burning with a characteristic *violet flame.* It will roll over the surface of the water (being lighter than that liquid), making a hissing noise, giving off white fumes, and disappear at last with a slight explosion. On this account it will be advisable to have a glass plate at hand to cover the dish, as a piece of the hot metal is sometimes thrown out.

Fig. 12.

The *potassium* tears asunder the *molecules* of **Water** with which it comes in contact, liberates a *part* of the **Hydrogen,** which formed one of its constituents, combining with the other part, and the remaining constituent, to form caustic potash or potassic hydrate. The latter may be seen as a white streak, dissolving in the water, to which it gives a peculiar taste, called *alkaline*. A *yellow turmeric paper*, which is not affected by ordinary water, will be turned *brown* upon being dipped into that in which the potassic hydrate is dissolved. A *reddened litmus paper* will be turned *blue*. This is called a *basic* or *alkaline* reaction.

*Exp.* 2. Take a piece of the metal *sodium*, it is cheaper than *potassium*, being one of the constituents of common salt. It is soft, like potassium, and lighter than water; but its action is not so *violent*. Throw a piece upon water, it will roll about, but will not inflame unless the water be *hot*, or its motion be arrested by putting it on a bit of blotting-paper; then it will do so, burning with a *golden yellow flame*. Cover with a glass plate before the flame is extinguished, as in the previous experiment.

Caustic soda, or sodic hydrate, will be formed in the same way as potassic hydrate was, *hydrogen* being *set free*; the same taste and basic reaction being also produced in this case.

*Exp.* 3. Fill a small cylinder or large test-tube with water, cover the mouth with a greased glass plate, and invert it in a basin of water. Wrap a bit of sodium in fine wire gauze, and throw it into water, it will sink on account of the weight of the latter. Bring the cylinder over it, *bubbles of gas* will rise and displace the water, speedily filling the tube. Cover with the thumb, or a greased glass plate, and remove from the water. The gas is transparent and colourless, and inflames when a light is applied, burning with a pale lambent flame. It is *hydrogen*, which has been liberated from the *water*.

Fig. 13.

Another, and perhaps a better way, is to get about an inch of *tin pipe* (such as used by gasfitters). Perforate it with holes, and then put a bit of *sodium* into it. Flatten the ends of the pipe to close them, and drop it into the water. Streams of **Hydrogen** will rise from the holes where the water comes in contact with the metal. The *sodium* decomposes the water, as in the case of the *potassium*.

Potassium must not be used, as the heat produced by the violent chemical action would inflame the gas. Care is necessary, even with *sodium*.

A pair of brass tongs having a bowl something like a bullet-mould, but perforated with holes, is very handy for this experiment, as they can be brought under a cylinder readily, and do not let the sodium escape and take fire, as it sometimes does, inflaming the hydrogen. They must be wiped dry if a second piece of metal has to be used. A wire-gauze spoon may be inverted over a bit of the metal as it floats on the surface of the water, and thus be brought under the cylinder.

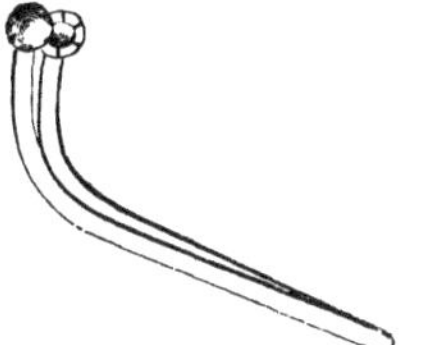
Fig. 14.

Water, however, is not the only liquid from which hydrogen can be obtained by means of the alkali metals, potassium and sodium. Two liquids, hydrochloric acid and ammonia, having very opposite properties with regard to their action on test-papers, yield this gas under certain conditions.

**43. Hydrochloric acid** is a gas, very soluble in water, its solution being kept in the laboratory under the same name. A *blue litmus* paper is instantly turned *red* when held over the open mouth of a bottle containing it.

**44. Ammonia** is also a gas, and its solution in water is kept by the chemist as a common reagent. It has the property of turning a *reddened litmus* paper *blue* and a *turmeric* paper *brown*, and is sometimes called the *volatile alkali*, to distinguish it from the *fixed alkalies, potash*, and *soda*.

These two liquids give off their respective gases when heated, as they are held simply in solution, and are more volatile than the water in which they are dissolved.

*Exp.* 4. Fit up a 4-ounce flask with a thistle funnel, and a delivery-tube bent twice at right angles, as in Fig. 15. The funnel-tube must go nearly to the bottom of the flask, the delivery-tube just passing through the cork. The longer leg of the latter must pass through the cork of a rather wide-mouthed bottle nearly to the bottom, and from the top of this bottle another elbow-tube is to be fitted, to which a piece of hard glass tube is to be attached, or a bulb-tube, if one be at hand.

Pour into the *flask* (*a*) a small quantity of strong solution of commercial *hydrochloric acid*, and half fill a bottle with pieces of pumice stone, saturated with strong sulphuric acid (oil of vitriol). Put a piece of potassium into the bulb of the tube (*c*). Heat the hydrochloric acid gently in the flask over a sand-bath;

the gas charged with moisture will pass out into the bottle, where it will be robbed of its moisture by the sulphuric acid spread over the large surfaces of pumice (dessicating agent), and the gas will pass off *dry* through the bulb-tube (*c*); or pass the moist gas through a bent tube containing chloride of calcium as

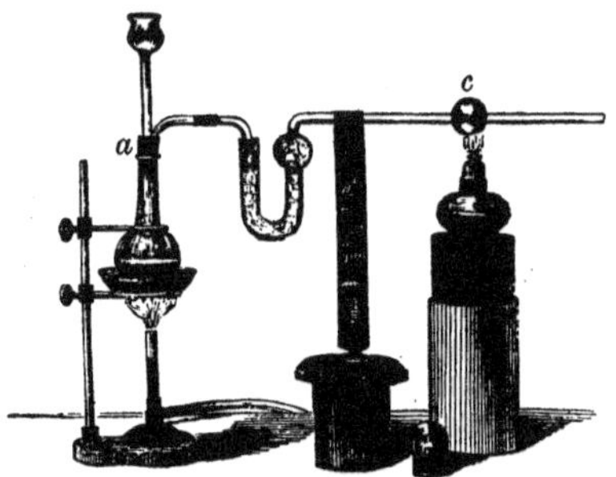


Fig. 15.

in Fig. 15, the gas will be dried. It will redden a blue litmus paper held at the open end, and fumes in air, dissolving in its moisture.

Now, heat the potassium in the bulb (*c*), by means of a spirit lamp. Chemical action will be set up, and the metal will glow. The hydrochloric acid gas will be decomposed, one of its constituents—chlorine—combining with the potassium to form potassic chloride, the other being set free.

Apply a light to the end of the tube; a gas inflames, which is *hydrogen*, the other constituent of hydrochloric acid.

*Exp.* 5. In the same manner, and with the same arrangement of apparatus, *ammonia gas* may be proved to contain hydrogen, the other constituent combining with the metal. The gas obtained by heating the solution must be dried, however, by passing it through quick-lime as a dessicating agent, as sulphuric acid forms a compound with ammonia. Chloride of calcium in a U tube might be used in both cases for drying the gases instead of the above. A test-tube containing sodium amalgam—made by putting bits of the metal into mercury, slightly warmed in an evaporating dish—may be used instead of potassium in both experiments, and heat need not be applied.

We have seen, therefore, that water, hydrochloric acid, and ammonia, all contain hydrogen as one of their constituents, and this has been separated from them by means of potassium and sodium, which have a great affinity for the other constituents contained in these compounds. It is only possible, however, by this means, to separate one constituent.

We shall, in the next chapter, proceed to show the nature of the other constituents contained in these bodies.

## SUMMARY OF CHAPTER V.

Water was called an element by the ancients.

The attraction of cohesion is loosened by heat. The heat of a series of electric sparks decomposes steam.

Potassium burns with a violet flame. Sodium burns with a golden yellow flame.

Both metals are soft, and can be cut with a knife. They are both lighter than water. Take care of sudden burst of steam. The metals move about on a cushion of steam.

Potassium and sodium tear asunder the molecules of water, hydrochloric acid, and ammonia—decompose them, in other words—liberating part of the hydrogen in water, combining with the other part and the other constituent to form potassic and sodic hydrates, which dissolve in the water. These bodies have a basic or alkaline reaction on test-papers, turning turmeric *brown* and reddened litmus *blue*. They give the water a peculiar taste, and make it slimy to the touch.

Sodium is not so violent in its action as potassium.

Sodium amalgam may be substituted for potassium.

Hydrogen gas can be collected over water by bringing a bit of sodium below an inverted vessel filled with water, the hydrogen replacing that liquid on account of the greater lightness of *the* gas.

Hydrogen is a transparent, colourless gas, without taste or smell. It is inflammable, and burns quietly down the jar.

Potassic, or sodic *chloride* or *nitride*, is formed when hydrochloric acid or ammonia gas is passed over these metals heated in a bulb-tube.

## CHAPTER VI.

DECOMPOSITION OF HYDROCHLORIC ACID, WATER, AND AMMONIA, BY THE ELECTRIC CURRENT—CONSTITUENTS OF THESE LIQUIDS—ANALYSIS AND SYNTHESIS—ELEMENTS AND COMPOUNDS—CHLOROUS AND BASYLOUS ELEMENTS.

45. FOR the experiments in this chapter, it will be necessary to have some means of generating an electric current by chemical action—that is, we must have some form of galvanic battery. Grove's, Bunsen's, or Smee's are the best. It is advisable to have four or five cells,* but two 4-inch of Bunsen's or Grove's arrangement will enable the teacher or student to perform the necessary experiments on a small scale, if also provided with a few other bits of apparatus, to be mentioned as we proceed. One cell is insufficient to produce electric decomposition, as it is unable to overcome the resistance a liquid opposes to the passage of the current.

*Exp.* 1. *a.* Get two pieces of platinum foil, and fasten a copper, or better, a platinum wire to each by making two pin-holes in the slips, and clamping the wires. Pour some *hydrochloric acid* into a beaker, and having set two cells of your battery in action, connect one slip of platinum with each pole. Dip the plates into the acid, bubbles of gas will be seen to arise from each of them, and a pungent, suffocating smell will be experienced. If a moistened litmus paper be held over the beaker it will be *bleached.*

*b.* Wash the plates well, and then dip them into another beaker containing *water*. A similar result will take place, bubbles of gas rising from each plate, but this time there will be no pungent smell.

*c.* Thirdly, dip the plates into a beaker containing a strong solution of ammonia, bubbles of gas will arise also in this instance.

---

* *Five* 4-*inch* cells of Bunsen's arrangement require 14 oz. by weight, or about 10 fluid oz. (½ a pint), of strong nitric acid to charge the porous cells; and 1 pint of dilute sulphuric acid (1 acid to 6 water) to charge outer cells containing amalgamated zinc plates. The nitric acid should be renewed every third or fourth time it is used, new dilute sulphuric acid every other time.

**In all three cases the liquids are decomposed by the electric current, which passes through the liquid from plate to plate.**

*Exp.* 2. Get a U or V shaped tube, or make one. Fit it to a stand by clamping it with a piece of tin or with some plaster of Paris. Fit two good corks into it, making a small hole through

Fig. 16.

each, and pass the wires attached to the platinum plates used in the previous experiments through them, so that the plates will dip into the tubes, as in Fig. 16. Half fill the tube with hydrochloric acid. Put the corks in tightly and connect the wires with the battery. The gases arising from each plate, as in the previous experiments, will collect now in the two limbs of the tube separately (of course mixed with air in this arrangement).

When the operation has been going on—say for five minutes—uncork the tube connected with the negative (—) or zinc end of the battery, a colourless gas, which inflames when a light is applied, is found to be present, and will be easily recognised as **hydrogen**. Remove the cork from the other limb, in which the plate connected with the positive (+) pole was put, the pungent smell spoken of before will be perceived. If a moistened litmus or turmeric paper be held over the tube, it will be bleached; or, if some solution of sulphate of indigo be previously introduced into this limb, it will be decolourised. The gas, which has a greenish yellow colour, is **chlorine**, and this bleaching property is one of its chief characteristics.

**46. Hence we learn that hydrochloric acid consists of two gases—hydrogen and chlorine.** Exact experiments moreover show that these gases are liberated from that liquid in equal volumes. These are, however, somewhat too advanced to be introduced or described at this stage. Chlorine is rather soluble in water, hydrogen only to a very small extent.

In the last chapter it was shown that the metal potassium turned out one of the constituents from hydrochloric acid, water, etc., but the *electric* current sets them *both* free. It in fact tears asunder each molecule of the acid, which consists of an atom of hydrogen combined with an atom of chlorine.

*Exp.* 3. Get a small voltameter, consisting of a receiver fixed to a stand, having two slips of platinum foil cemented into the bottom, connected by copper wires with two binding screws. Fit it with a cork and delivery-tube, as in the figure.

Fig. 17.

Fill the apparatus with water, to which a small quantity of sulphuric acid has been added, to increase its conducting power. Send the current from the battery through it; two gases will stream off at the platinum plates, which may be collected over water in a *soda-water bottle* or *stout glass tube*. The electric current in passing through the water tears its molecules asunder, and the resulting gases mix together in the bottle.

Cover the mouth with a greased glass plate, and remove the bottle from the water. Wrap it round with a duster, to protect the hand in case of fracture. Remove the greased plate and get some one to apply a light; a loud explosion will take place, caused by the rushing together of the atoms of the mixed gases to form *water gas*, which speedily condenses to water on the sides of the bottle.

A substitute for the voltameter above described may be made by fitting the two *platinum plates* to a cork in a wide-mouthed bottle, the cork being also furnished with a delivery tube. If, however, the current be strong, the exposed platinum is apt to become red-hot and fire the mixture, blowing out the cork.

*Exp.* 4. Make some soap-suds in a basin, and let the delivery-tube from the voltameter in the last experiment pass into it and blow some bubbles. Take some of these in the palm of the hand. Stand away from the voltameter, and apply a light, the bubbles will burst with a loud report, but of course will do no injury to the hand. Or, *remove the delivery-tube*, and apply a light to the bubbles in the basin.

A small collodion balloon may be filled with the mixed gases and exploded as it ascends by attaching a piece of string soaked in strong solution of *nitre*, with a bit of lighted gun-cotton at the end.

**47. In all these cases chemical combination takes place, WATER being formed again from the constituents.**

Let us now, as in the case of hydrochloric acid, collect the two gases separately. This we may do by means of the same arrangement with a U tube, but it is better done in the following manner: Get two holes drilled in the bottom of a glass basin, and cement two pieces of platinum foil in the bottom, with copper wires attached, passing through the holes. Two equal-sized test-tubes are to be inverted over the platinum plates. A very neat and not expensive instrument of this kind is shown in Fig. 18. It is also called a voltameter, and is used for collecting the gases separately.

*Exp.* 5. Fill the tubes and half fill the basin with water, adding some sulphuric acid to both to increase the conducting power. Invert the tubes over the slips of platinum and connect with the battery. The current in passing from plate to plate will decompose the water, the gas rising in the tube over the plate connected with the negative pole being double in quantity to that which rises over the plate connected with the positive pole.

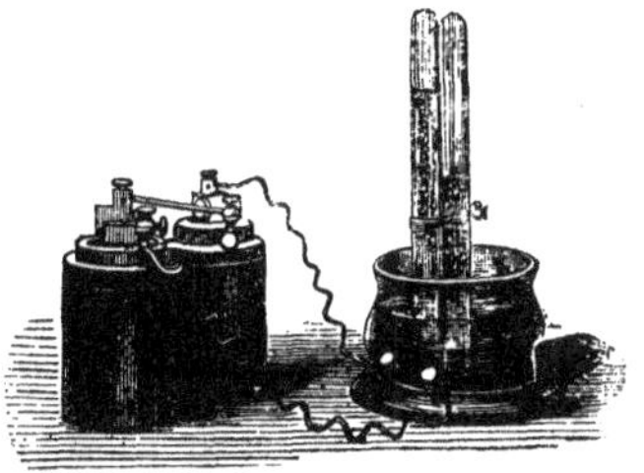

Fig. 18.

Stopping the current when a small quantity of gas has collected, removing the tube over the *negative* pole by means of a glass plate, and applying a light, it will be found to be **hydrogen.** Removing the other tube and plunging a glowing splinter into the gas, it will be rekindled, with a slight pop, if the tube have much gas in it. This gas is *oxygen,* the other constituent of the water.

**48. In this case we not only learn that water consists of two gases—oxygen and hydrogen—but we find they are**

**liberated in the proportion of one volume of the former to two of the latter.**

*Exp.* 6. Make the same arrangement with a U tube as in the case of hydrochloric acid. Pour in some strong ammonia solution, and *add cautiously* a few drops of sulphuric acid, to increase its conducting power. Connect the plates with the battery; hydrogen again appears at the negative pole, and may be tested in the usual manner. On removing the cork from the other tube and applying a light, it will be extinguished by a small quantity of gas which has collected there. This gas is *nitrogen*, the other constituent of ammonia.

**49. Exact experiments show that three volumes of hydrogen are combined with one volume of nitrogen in ammonia.**

We have thus shown by **analysis** the constitution of the three compounds—hydrochloric acid, water, and ammonia, and it now remains for us to prove as far as possible, by putting the constituents of these substances together (that is, by synthesis), that these compounds are formed again.

To do this the following experiments must be performed: fuller particulars for preparing and collecting the gases will be found on pages 66 and 67.

*Exp.* 7. Put some zinc clippings into a Woulff's bottle, fitted with a thistle funnel and delivery tube. Pour on them some common hydrochloric acid. The acid will be decomposed, the hydrogen being liberated. Collect a jar of the gas over water at the pneumatic trough, and cover it with a greased glass plate.

Chlorine may be obtained from hydrochloric acid by the action of *black oxide of manganese*, or manganic dioxide.

*Exp.* 8. Put into a 4-oz. flask (or Florence flask) some of this manganic dioxide, and pour on it some hydrochloric acid, carefully mixing it, so that all the powder shall be wetted and made into a thin paste. Fit a cork and delivery-tube to the flask, and let the latter pass to the *bottom of a jar* as in Fig. 19, the jar being of the same size as that in the previous *Exp.* Cover the jar with a card having a hole in the middle or a notch at the side. Heat the flask gently over a sand-bath. Chlorine gas will come off and lift the air out of the jar on account of its greater density. Great care must be taken not to let any escape into the air, as it is poisonous, unless largely mixed with air (*see* Chap. IX.). Apply a moistened test-paper to the mouth of the jar from time to time, and when this is bleached the jar may be considered full.

*Exp.* 9. Cover the jar with a greased plate, and invert it over the one containing the hydrogen. Holding the jars firmly, draw the two plates from between them. The two gases will mix together, the heavier gas descending and the lighter ascending—diffusing, as it is called. This may be assisted by turning them up and down once or twice. The gases, however, only mix, they do not combine.

Fig. 19.

Light a spirit lamp, and holding a cylinder in each hand, bring them to the flame and separate them as in Fig. 21. An explosion will take place, and dense white clouds will be produced, which *redden* litmus paper. They consist of hydrochloric acid, showing conclusively that hydrogen and chlorine are the **only** constituents of hydrochloric acid. They combine, moreover, in equal volumes; one volume of hydrogen combining with one volume of chlorine to form two volumes of hydrochloric acid gas. The relative weights of these vols. being as 1 to 35·5, chlorine being 35½ times the weight of hydrogen. The weight of the product is the sum of the two.

Fig. 20.

## To prove the composition of water synthetically.

*Exp.* 10. Take the stout tube (explosion tube) used in a former *Exp.*, or the soda-water bottle, and fill with water. Ascertain its measure in fluid ounces or in any other way. Divide its volume into three parts, and pour one-third part of the water into the tube, pasting a piece of paper or putting an india-rubber band round to mark the height. Pour in another third of the water, and mark its position by another band. One-third of the tube will be left. This is *calibrating* a tube.

Fig. 21.

Now, fill the tube with water and invert it in the trough. Fill it two-thirds with hydrogen and one-third with oxygen, which is to be prepared thus:

*Exp.* 11. Take a small test-tube of *hard* glass. Fit it with a cork and delivery-tube, and put in it some oxygen mixture (*see* Chap. XI.). Heat in the Bunsen flame; a colourless gas will be given off, which is to be collected in the above tube at the pneumatic trough.

The explosion tube being now filled with a mixture of ⅔ hydrogen and ⅓ oxygen, cover it with a greased plate and remove it from the trough. Bring the open mouth of the tube to the flame of the spirit lamp, a loud explosion will be the result, and water is formed. If more or less than ⅔ of hydrogen, or more or less than ⅓ of oxygen be present, the explosion will not be so loud, and some of the gas in excess will be left free.

*Note.*—A small test-tube of oxygen may be collected first. The gas will rekindle a glowing splinter.

**50. In this case two volumes of hydrogen combine with one volume of oxygen to form water.**

Exact experiments to be mentioned hereafter prove also that the two volumes of hydrogen, combining with one of oxygen, form only *two* volumes of water vapour—that condensation thus takes place.

The weights of oxygen and hydrogen entering into combination are as 16 to 2—oxygen being 16 times the weight of the same volume of hydrogen. The product of the combination is the sum—18 parts by weight.

**51. Now, there is no method known by which hydrogen can be made to combine directly with nitrogen, so that the composition of ammonia cannot be proved SYNTHETICALLY, as in the other two cases.** But in any given quantity of ammonia gas, it is found that for every 3 parts by weight of hydrogen liberated, there are 14 parts by weight of nitrogen. And, moreover, if electric sparks be sent through ammonia gas in the spark-stream apparatus, the gas is decomposed, and its volume becomes **doubled**. That is, 2 volumes of ammonia gas become 4 volumes on decomposition; hence we infer that 3 volumes of hydrogen, and 1 volume of nitrogen (weighing 14 compared with hydrogen as unity) are condensed to two volumes of ammonia gas.

We have in this way become acquainted with certain elements contained in the three compounds, *hydrochloric acid*, *water*, and *ammonia*, and have learned by the way some of their properties. These compounds are types of many others built upon the same plan.

It was found in the electric decomposition of these compounds, that **hydrogen** always appeared at the negative pole; on this account it is called an **electro-positive** element (oppositely electrified bodies always attracting each other), the term **basylous** being sometimes applied. On the other hand, the elements **chlorine, oxygen,** and **nitrogen,** appeared at the positive pole; and are hence called **electro-negative** elements—the term **chlorous** being frequently applied to them from the first named.

The elements are therefore sometimes divided into two groups according to the way they behave when submitted to the electric current.

Thus, the following elements are chlorous or electro-negative to the others, which are more or less electro-positive or basylous:—Fluorine, chlorine, bromine, iodine, oxygen, sulphur. Chemical affinity is greatest between a chlorous and a basylous element, but combination takes place at times between two of the chlorous or between two of the basylous class. Mercury, for instance, is chlorous to sodium, but basylous to chlorine; the second compound is, however, much more stable (not so easily decomposed) than the first.

Fluorine is more negative than chlorine; chlorine more negative than bromine; and bromine than iodine. Hence, if bromine be added to potassic iodide, the iodine is liberated and potassic bromide formed. If chlorine be passed into a solution of potassic bromide, bromine is liberated and potassic chloride formed. In the same way oxygen is more negative than sulphur, and so on.

A table of the most common elements will be given in the next chapter.

## CHAPTER VII.

Combination by Volume and Weight—Table of the Elements Atomicity of the Elements — Compound Radicals — Chemical Nomenclature—Acids, Bases, and Salts—Chemical Equations.

**52.** As this book is intended to furnish the student principally with the **practical** details of the subject, the writer would refer to the numerous works already published for fuller information on the **theoretical** part, as space cannot be allotted to both.

Something, however, must be said about the nomenclature used, and the modes of expressing the composition of bodies, and their reactions upon one another, in order to avoid too frequent explanation of similar matters.

It has been shown in the previous chapter, by the electric decomposition of hydrochloric acid, water, and ammonia, that these bodies were built up of certain **elements** combined together in certain proportions by volume and weight. Thus :—

1 vol. of hydrogen combined with 1 vol. of chlorine forms **2 vols.** of hydrochloric acid gas.

2 vols. of hydrogen and 1 vol. of oxygen,—**2 vols.** of water gas.

3 vols. of hydrogen and 1 vol. of nitrogen, form **2 vols.** of ammonia gas.

Taking the weight of a litre of hydrogen as unity = 1, a litre of chlorine weighs 35·5 ($35\frac{1}{2}$), a litre of oxygen 16, and a litre of nitrogen 14 times as much. But as gases *expand* by heat and under diminished pressure, and *contract* on cooling and under increased pressure, it is necessary to compare them at the *same temperature* and pressure. These weights are called the *volume weights* or specific gravities of these gases compared with hydrogen, and the same holds true of others.

Now the weights entering into combination in the **typical compounds** taken are :—In hydrochloric acid, 1 of hydrogen to 35·5 of chlorine; in water, 2 of hydrogen to 16 of oxygen; and in ammonia, 3 of hydrogen to 14 of nitrogen. To these may be added a *fourth type*, which consists of 4 parts by weight of hydrogen combined with 12 parts by weight of carbon, forming **marsh gas.**

The volumes and weights of the elements thus com bined form in all cases molecules, which measure the same as 2 volumes of hydrogen.

Thus chemical combination can be expressed by **volume** in cases where the elements can be obtained in the state of vapour or gas, and *in all cases* by **weight.**

Thus, to each element a number has been assigned, showing its relative weight compared with hydrogen as unity. This is supposed to be the weight of the "atom," and hence the term *atomic weight.*

It is usual for the chemist to use a set of **symbols** to represent the elements—a sort of chemical shorthand, in fact. The initial letter of the common, or sometimes that of the Latin name, is used for this purpose. This symbol also, by remembering or referring to the atomic weight of the substance, expresses the quantity by weight of the substance entering into combination. Thus O not only denotes oxygen, but an atom or 16 parts by weight of that element; H, an atom or 1 part by weight of hydrogen; Cl, an atom of chlorine, or 35·5 parts by weight; N, an atom, or 14 parts by weight of nitrogen; C, an atom, or 12 parts by weight of carbon. Two or more atoms of an element are expressed by a small figure placed below the symbol on the right, as $H_2$. The symbol also expresses 1 volume of the elements in the gaseous state.

When the names of two elements begin with the same letter, a second is added to that most recently discovered, as Cl. for chlorine, Br. for bromine, Pb. (Lat. *plumbum*) for lead; C. standing for carbon, B. for boron, and P. for phosphorus.

The names of the more important elements are printed in the following table in the largest type.

The symbols of the elements forming a compound are written side by side, with small figures as above, if more than one atom of either element is used, as $CH_4$, $Fe_2O_3$.

## TABLE OF MOST IMPORTANT ELEMENTS.

| Non-Metallic (*Metalloids*). | | | Metallic. | | | | | |
|---|---|---|---|---|---|---|---|---|
| | Symbol and Atomicity. | Atomic Weights. | | Symbol and Atomicity. | Atomic Weights. | | Symbol and Atomicity. | Atomic Weights. |
| **Hydrogen**, . | $H'$ | 1 | **Aluminium**, . . . | $Al^{iv}$ | 27·5 | **Magnesium**, . . . . | $Mg''$ | 24 |
| **Chlorine**, . | $Cl'$ | 35·5 | Antimony (Stibium), | $Sb^{v}$ | 122 | **Manganese**, . . . . | $Mn^{vi}$ | 55 |
| Bromine, . | $Br'$ | 80 | Arsenic, . . . . . | $As^{v}$ | 75 | **Mercury** (Hydrargyrum), | $Hg''$ | 200 |
| Iodine, . . | $I'$ | 127 | **Barium**, . . . . | $Ba''$ | 137 | Nickel, . . . . . . | $Ni^{vi}$ | 58·8 |
| Fluorine, . | $F'$ | 19 | Bismuth, . . . . | $Bi^{v}$ | 208 | Palladium, . . . . . | $Pd^{iv}$ | 106·5 |
| **Oxygen**, . . | $O''$ | 16 | Cadmium, . . . . | $Cd''$ | 112 | **Platinum**, . . . . . | $Pt^{iv}$ | 197·1 |
| **Boron**, . . | $B'''$ | 11 | **Calcium**, . . . . | $Ca''$ | 40 | **Potassium** (Kalium), . | $K'$ | 39 |
| **Carbon**, . . | $C^{iv}$ | 12 | Chromium, . . . | $Cr^{vi}$ | 52·5 | **Silver** (Argentum), . . | $Ag'$ | 108 |
| Silicon, . . | $Si^{iv}$ | 28·5 | Cobalt, . . . . . | $Co^{vi}$ | 58·8 | **Sodium** (Natrium), . . | $Na'$ | 23 |
| **Nitrogen**, . | $N^{v}$ | 14 | **Copper** (Cuprum), . | $Cu''$ | 63·5 | Strontium, . . . . . | $Sr''$ | 87·5 |
| **Sulphur**, . | $S^{vi}$ | 32 | Gold (Aurum), . . | $Au'''$ | 196·7 | Thallium, . . . . . | $T'$ | 204 |
| Selenium, . | $Se^{vi}$ | 79 | **Iron**, . . . . . | $Fe^{vi}$ | 56 | **Tin** (Stannum), . . . | $Sn^{iv}$ | 118 |
| **Phosphorus**, | $P^{v}$ | 31 | Lead (Plumbum), . | $Pb^{iv}$ | 207 | **Zinc**, . . . . . . . | $Zn''$ | 65 |
| | | | Lithium, . . . . | $Li'$ | 7 | | | |

## TABLE

*Showing Combination by Volume of*

TYPICAL COMPOUNDS.

| | | | | | | | | |
|---|---|---|---|---|---|---|---|---|
| | | | H = 1 | + | Cl = 35·5 | = | HCl = 36·5 |
| | | H = 1 | H = 1 | + | O = 16 | = | $OH_2$ = 18 |
| | H = 1 | H = 1 | H = 1 | + | N = 14 | = | $NH_3$ = 17 |
| H = 1 | H = 1 | H = 1 | H = 1 | + | C = 12 | = | $CH_4$ = 16 |

The symbols of a compound thus written are called its *formula*, and represent *one molecule*. Two or more molecules are shown by a figure on the left of the symbols, as 2HCl, $4OH_2$—the figure multiplying the atoms of all the elements in the molecule. Thus, in the first instance given, there would be two atoms of hydrogen and two atoms of chlorine combined, forming two molecules of hydrochloric acid; whilst in the second case there would be four atoms of oxygen, and four times two, or eight atoms, of hydrogen, representing four molecules of water.

As a means of further illustrating this subject, the combination by volume to form the *typical compounds* before mentioned is given on page 59 here, as represented pictorially by Dr Hofmann.*

The weight of the double volumes (molecules of compound gases) being divided by 2, gives, of course, the single volume weight of the respective compounds. Thus :—

$$\overset{\text{H} \quad \text{Cl}}{\frac{1 + 35{\cdot}5}{2}} = \frac{36{\cdot}5}{2} = \overset{\text{Volume wt. of HCl}}{18{\cdot}25}\,; \quad \overset{\text{H} \quad \text{O}}{\frac{2 + 16}{2}} = \frac{18}{2} = \overset{\text{Vol. wt. Water Gas.}}{9}\,;$$

$$\overset{\text{H} \quad \text{N}}{\frac{3 + 14}{2}} = \frac{17}{2} = \overset{\text{Vol. wt. of NH}}{8{\cdot}5}\,; \text{ and } \overset{\text{H} \quad \text{C}}{\frac{4 + 12}{2}} = \frac{16}{2} = \overset{\text{Vol. wt. of Marsh Gas.}}{8{\cdot}}$$

It will be seen by the examples already adduced, that the elements Cl, O, N, and C have very different powers of entering into combination with hydrogen, and so it is with other elements. One atom of Cl can only combine with one atom of H, whilst an atom of O can combine with two, N with three, and C with four atoms of the same element. Hence an atom of O is able to replace, or is *equivalent* to, *two* atoms of Cl; an atom of N to *three*, and an atom of C to *four* of chlorine or hydrogen. And so of other elements: an atom of potas-

* A series of card-board boxes can be easily obtained to fit into one another (a nest), to show this combination by volume, and the condensation in all cases (but the first) to the double unit-volume.

sium (K), sodium (Na.), or silver (Ag.), combines with a single atom of chlorine to form KCl, NaCl, or AgCl—potassic, sodic, or argentic *chlorides;* but it requires two atoms of these metals to combine with one of oxygen to form potassic, sodic, or argentic *oxides.* Again, the metals magnesium, bismuth, platinum, and antimony require *two, three, four,* and *five* atoms respectively to form $MgCl_2$, $BiCl_3$, $PtCl_4$, and $SbCl_5$, the chlorides of these metals.

This difference of combining power, the knowledge of which is so useful to the chemist, is sometimes called the "atomicity," "equivalence," or "quantivalence" of the element, which may be defined to be *its power of displacing, or entering into combination with, another element.*

Those elements whose atoms are equivalent to one atom of hydrogen are called *monads;* those whose atoms can replace two atoms of H, *dyads;* those whose atoms can replace three, *triads;* four, *tetrads;* five, *pentads;* and six, *hexads;*—as in the following table:—

ATOMICITY OF THE COMMON ELEMENTS.

| MONADS. | DYADS. | TRIADS. | TETRADS. | PENTADS. | HEXADS. |
|---|---|---|---|---|---|
| Hydrogen. | Oxygen. | Boron. | Carbon. | Nitrogen. | Sulphur. |
| ——— | ——— | ——— | Silicon. | Phosphorus. | Selenium. |
| Chlorine. | Barium. | Gold. | ——— | ——— | ——— |
| Bromine. | Strontium. | ——— | Tin. | Arsenic. | Chromium. |
| Iodine. | Calcium. | | Aluminium. | Antimony. | Manganese. |
| Fluorine. | Magnesium. | | Platinum. | Bismuth. | Iron. |
| ——— | Zinc. | | Palladium. | ——— | Cobalt. |
| Potassium. | Cadmium. | | Lead. | | Nickel. |
| Sodium. | Mercury. | | ——— | | ——— |
| Lithium. | Copper. | | | | |
| Thallium. | ——— | | | | |
| Silver. | | | | | |
| ——— | | | | | |

The atomicity of an element is indicated when symbols only are used, by using dashes, or Roman numerals, on the right of the symbol, as in the "Table of Most Important Elements."

The atomicity of elements, and their modes of combining with others, may also be represented by their symbols having lines, bonds, or dashes standing out from them, at the other ends of which the symbols of the combining elements may be written. These are the graphic formulæ of Dr Frankland. Thus the atoms of H and Cl would be represented as having *one* bond, Zn and O with *two*, B and Au with *three*, C and Pt with *four*, P and N with *five*, and S with *six* dashes.

| | | |
|---|---|---|
| Chlorine, Cl— | Gold, — Au — (with one bond below) | Phosphorus, — P — (with bonds above and below) |
| Zinc, — Zn — | Carbon, — C — (with bonds above and below) | Sulphur, — S — (with bonds above and below) |

Potassic, magnesic, and auric chlorides would be represented thus:—

$$Cl - K \; ; \; Cl - Mg - Cl \; ; \; Cl - \overset{\overset{\large Cl}{|}}{Au} - Cl$$

Zincic oxide, Zn = O

Atoms when isolated unite with other atoms, and form themselves into molecules. A molecule must have all its bonds engaged, that is, it cannot combine with any substance without altering the arrangement of the atoms. Hence, there must always be an even number of *bonds* in the molecule of any element or in any compound.

Some elements with an even number of bonds can exist as *monatomic molecules*, their own bonds satisfying each other, as

Mercury Hg (/....|)   Zinc —Zn— (bonds joined by dotted arc)   and Cadmium Cd (/...\)

The molecules of most elements, however, are *diatomic*, that is, are made up of two atoms, as

H — H   Cl — Cl   O = O   N ≡ N

Oxygen in the state of **OZONE** has three atoms in its molecule, it is *triatomic*, O—O \O/ ; phosphorus and arsenic are *tetratomic;* and sulphur is *hexatomic,* or has six atoms in its molecule.

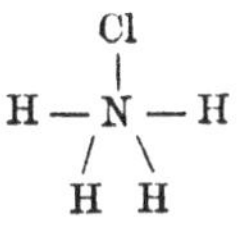

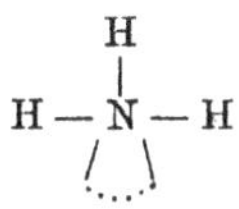

A pair of the bonds of an element may unite together or become latent. Thus nitrogen, which has *five* bonds, as in ammonic chloride ($NH_4Cl$), may have *three* bonds active and *two* latent, as in ammonia gas; or it may have only *one* active and *four* latent. Hence the terms "absolute," "latent," and "active," atomicity.

```
     Cl
     |
 H — N — H
    / \
   H   H

     H
     |
 H — N — H
    /  \
    ....
```

The bonds always diminish or increase in pairs. The element having the greatest number of bonds in a compound is usually placed first, and is printed in thick type in the following pages, as **N**$H_3$, **O**$H_2$.

Some elements in compounds group themselves together and act as elements. These *atom groups* are called *compound radicals,* as hydroxyl. $Ho_2$ the molecule, and Ho the atom or semi-molecule; as, in sulphuric or nitric acids, **S**$O_2Ho_2$ and $NO_2Ho$, the figure on the right multiplying the group (2 of H and 2 of O), the small *o* being written for quickness, and to show that it is part of the group. The same atom fixing power belongs to them as to the elements, they are hence monads, dyads, triads, etc., replacing and displacing one another just as elements do. Hydroxyl Ho is a monad radical, whilst the dyad radical zincoxyl is represented by **Zn**o″, with the two dashes to mark its atomicity. **Zn**o″, or its analogues, will replace $Ho_2$ in compounds, or $Ko_2$, etc.

**53. Nomenclature.**—The names given to the elements are for the most part arbitrary, but are easily remembered.

In *binary compounds,* that is those consisting of two elements, the root of the name of the *electro-positive or basylous* element comes first, being made to end in *ic,* followed by the root of the *chlorous or electro-negative* element, with the ending *ide,* thus:—

Sodium and chlorine form Sod*ic* chlor*ide* NaCl
Magnesium and oxygen form Magnes*ic* ox*ide* **Mg**O.
The same with sulph*ides,* brom*ides,* iod*ides,* nitr*ides,* etc.

If there be two compounds of the same elements, that which has the smaller quantity of the *chlorous* element takes the termination *ous* to first name, *ic* being used for that which has the larger quantity of the chlorous element, as

Mercur*ous* chloride, $Hg_2Cl_2$. Ferr*ous* sulphide, FeS.
Mercur*ic* chloride, $HgCl_2$. Ferr*ic* sulphide, $Fe_2S_3$.

Sometimes the prefixes *hypo* (under), and *per* (for hyper, more), are used to indicate a smaller or greater quantity of the chlorous element. In a few instances prefixes are used, indicating the number of atoms of each element, as triferric, tetroxide, $Fe_3O_4$, containing three atoms of iron and four atoms of oxygen.

Anhydrides are binary compounds, containing oxygen, which are converted into acids on the addition of water.

Sulphur*ous* anhydride, $SO_2$, gives sulphurous acid, $SOHo_2$.
Phosphor*ic* anhydride, $P_2O_5$, gives phosphoric acid, $POHo_3$.

**54. Acids** are compounds containing one or more atoms of hydrogen, which can be replaced by a metal or *metallic radical*, forming two kinds of *salts*. Examples of such combinations with metals, oxides, and hydrates, will be found throughout the book.

Thus, hydrochloric acid or hydric chloride, HCl, when added to a metal, has its H exchanged for an equivalent of that metal, forming what has been called a haloid salt, as sodic, zincic, boric, and platinic chlorides, NaCl, $ZnCl_2$, $BCl_3$, $PtCl_4$, which are binary compounds.

Acids containing oxygen form a class of salts called oxy-salts, which are ternary compounds, or contain three elements. Thus, sulphuric acid, $SO_2Ho_2$, added to zinc, forms zinc*ic* sulph*ate*, $SO_2Zno''$, where the radical hydroxyl, $Ho_2$, is exchanged for zincoxyl, **Zno**$''$. Those acids whose names end in *ic* form salts ending in *ate*; those ending in *ous* form salts ending in *ite*.

Acids are either *mono*basic, *di*basic, *tri*basic, or *tetra*-basic, as they have one or more atoms of replaceable hydrogen or hydroxyl, the term *polybasic* being frequently used for all but the first.

Nitric acid, $NO_2Ho$ (Monobasic), { only one class of salts, $NO_2Nao$, $NO_2Ago$.

Sulphuric acid, $SO_2Ho_2$ (Dibasic), { two classes, $SO_2Nao_2$, or $SO_2Zno''$; and $SO_2HoNao$.

Phosphoric acid, $POHo_3$ (Tribasic), { three classes, $PONao_3$, $PoHoNao_2$, $PoHo_2Nao$.

The use of symbolic and graphic formulæ enables the student to express the changes which take place in chemical actions by **Equations.** Thus, when hydrochloric acid was passed over heated potassium in the bulb-tube in *Exp.* 4, Chap. V., the reactions could be expressed in equations, graphically or symbolically, thus :—

Graphic equation : $\begin{matrix} H-Cl \\ H-Cl \end{matrix} + K-K = \begin{matrix} K-Cl \\ K-Cl \end{matrix} + H-H$

Symbolic equation : $2HCl + K_2 = 2KCl + H_2$

Where the two molecules of HCl, and one molecule or two atoms of potassium on the left, are represented by the same number of atoms of the elements on the right of the *sign* = which show the change that has taken place. Abundant instances of chemical equations will be given throughout the book.

---

## CHAPTER VIII.

### HYDROGEN H'.

**55.** WE shall now take up the study of certain typical elements, and their most important compounds, and for this purpose we shall commence with the *monad* element, hydrogen.

It will be recollected that we obtained this gas from hydrochloric acid (HCl), water $OH_2$, and ammonia $NH_3$, which were decomposed by the alkali metals, potassium (K), and sodium (Na), on account of the affinity of these metals for the other constituent in these compounds.

Now the metals zinc, iron, and aluminium, have a great affinity for chlorine, and are able, at ordinary temperatures, to decompose a solution of hydrochloric acid with evolution of hydrogen.

*Exp.* 1. Put a few grains of granulated zinc into a test-tube, and pour some common hydrochloric acid upon it; violent chemical action will take place, which increases on the addition of water. Hydrogen will be evolved, and will inflame on a light being applied to the mouth of the tube, whilst zincic chloride will be formed and dissolve in the water, from which it can be recovered by evaporation.

Zinc being a dyad,* an atom of the metal replaces two atoms of hydrogen, thus requiring two atoms of chlorine to combine with to form *zincic chloride*, the reactions being shown by the following equations:—

$$\text{Graphic: } \begin{matrix} H-Cl \\ H-Cl \end{matrix} + Zn = Cl - Zn - Cl + H - H$$

$$\text{Symbolic: } 2\,HCl + Zn = Zn\,Cl_2 + H_2$$

Iron and zinc are scarcely acted upon by concentrated sulphuric acid (common oil of vitroil), but are violently attacked when water is added, and dissolve with evolution of hydrogen.

*Exp.* 2. Pour some strong sulphuric acid on some scraps of zinc in a test-tube, little or no chemical action will take place. Add some water, the action will become violent, and hydrogen will be evolved.

Whence comes the hydrogen? Not from the zinc certainly, for that is an element; not from the water, for water is not decomposed by zinc alone. It must therefore have come from the acid. Now sulphuric acid is made up of three elements, sulphur, oxygen, and hydrogen; or, according to Dr Frankland's view, of two radicals, $SO_2$ and $Ho_2$ (hydroxyl), its formula being $H_2\,SO_4$,

* Zinc is also a *monatomic molecule*, hence the single atom can be used. Two atoms of all other elements except Zn, Hg, and Cd, must be used in equations, that is not less than a molecule, or $K_2$, $Na_2$, $H_2$, etc.

or, according to Dr Frankland's notation, $SO_2Ho_2$. In the former case, the atom of zinc is supposed to replace the two atoms of H, as in the decomposition of hydrochloric acid. In the latter, according to Dr Frankland's view, a new body is formed, zincoxyl (Zno″ a dyad radical), which displaces the two atoms or molecule of hydroxyl ($Ho_2$) in the sulphuric acid, as in the following equations:—

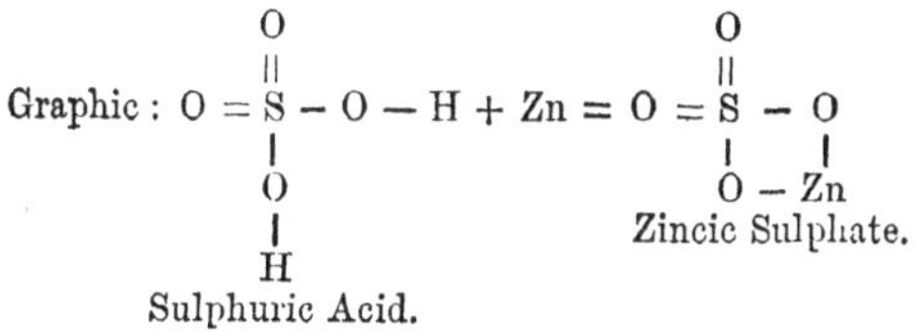

```
             O                      O
             ||                     ||
Graphic : O = S - O - H + Zn = O = S - O
             |                      |   |
             O                      O - Zn
             |                 Zincic Sulphate.
             H
      Sulphuric Acid.
```

Symbolic : $SO_2Ho_2 + Zn = SO_2\ Zno'' + H_2$

The water is required to dissolve the zincic sulphate as fast as it is formed, as this substance covers the metal directly, and prevents the further decomposition of the acid before water is added. In the case of iron, the sulphate interferes with the action considerably.

**56.** Now, let us examine the properties of this gas more fully. For this purpose we shall require several jars, and we may use either HCl or $SO_2Ho_2$, and Zn or (Fe). Hydrochloric acid is, however, preferable to sulphuric, and zinc to iron; the hydrogen thus obtained being purer and more rapidly evolved.

*Exp.* 3. Into a two-necked Woulff's bottle put some granulated zinc, and just cover it with water. Fit the bottle with a thistle-headed funnel and delivery-tube. Stop the delivery-tube with the finger, and blow through the funnel to see if it be *air-tight*. If considerable resistance is felt, and the air rushes out violently from the delivery-tube when the finger is taken away, the fittings may be considered sound. If, however, the air whistles through somewhere, the apparatus is not safe, the fault must be found and rectified before proceeding, and *only then may the student proceed safely*.

*Note.*—So many explosions are produced for want of attention to these points by young chemists, that the author must be excused if he has been too prolix. Hydrogen is *not* an explosive gas, but *it is so if mixed with air.*

Fill a pneumatic trough so that the water comes just above the shelf. Fill the jars with water in the trough and invert them—taking care they are *quite* full, standing two on the shelves over the holes. Have some earthenware trays, small plates, or saucers ready, and greased glass plates. Arrange the gas generating bottle so that the delivery-tube can be brought under the funnels in the shelf, as in the fig. Have also a small wax taper fixed to a wire ready to light when wanted.

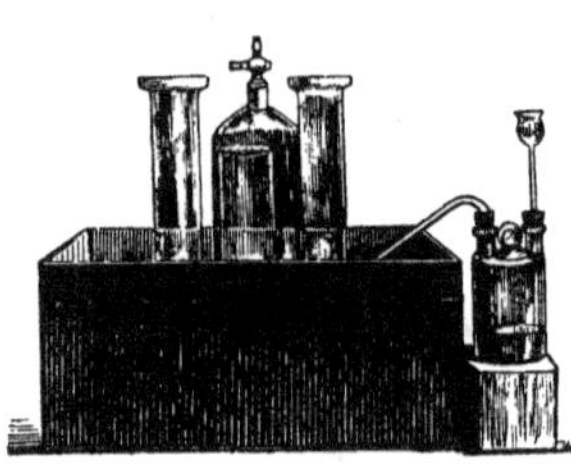

Fig. 22.

Pour some hydrochloric acid down the thistle-funnel, air will rush out of the delivery-tube and displace the water in the bottle. A violent effervescence will take place, and hydrogen gas *mixed with the air in the bottle* will pass over. *The first two or three jars of this mixture must therefore be thrown away.* After a short time, fill a test-tube with water, invert it over the hole in the shelf and fill it with gas. Bring it to a light, which should be burning far enough away from the generating flask. If it inflames with a *pop*, air is still mixed with the gas. Try another, and if necessary, a third test-tube, till the hydrogen burns with a pale, lambent flame, descending quietly into the tube. Then, and not till then, collect a jar of the gas, and when full, insert a tray or plate under it and remove it out of the way, putting another in its place. The water in the tray will prevent the escape of the gas. Cover others with greased glass plates.

If necessary, pour some more HCl into the generating bottle; but if the zinc be exhausted, the apparatus must be disconnected to add more, and the gas must be tested again to see if it be free from admixture with air before collecting in jars.

Having collected several cylinders or jars of the gas, try the following:—

*Exp.* 4. Place a cylinder of the gas on the table mouth upwards. Invert another jar, holding it with the left hand. Remove the greased plates from both jars, and after about half-a-minute bring a lighted taper to the first—the gas will be found to have escaped. Bring the light to the inverted jar, and the gas will inflame. Plunge the taper into the gas as in the figure, it will be extinguished, but can be rekindled at the mouth of the jar. This can be repeated till the gas is consumed.

The hydrogen escaped from the first jar on account of its lightness, whilst a jar for the same reason may be kept in an inverted position for a quarter of an hour before it has entirely dispersed. Its lightness may be further demonstrated by the following experiments.

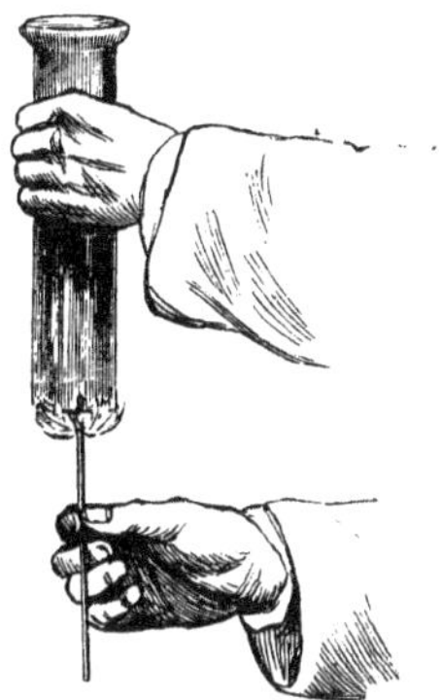

Fig. 23.

*Exp.* 5. Take two equal-sized cylinders, fill one with hydrogen and hold the other in an inverted position with the left hand. Carefully raise the jar containing the gas mouth downwards with the right hand, bringing it to the mouth of the other jar. Incline it as in the figure, and pour the gas upwards ; it will rise on account of its levity through the air and displace it. Stand the jar which previously contained the gas on the table, apply a light, the H. will be found to have left it, if the experiment has been well performed. Bring the light to the inverted jar, the gas will inflame with a slight report on account of mixture with air. The jar being open, no injury is likely to arise.

Fig. 24.

Explosions are only dangerous when the gases are confined, so that when expanded by heat they cannot easily get out. Hence bottles are not so good as cylinders for such experiments, and thin glass vessels should not be used.

*Exp.* 6. Have a jar furnished with a stop-cock or fitted with a cork, through which passes a piece of glass tube which may be stopped with a cork on the top till wanted. Fill the jar with gas at a rather deep pneumatic trough, or in a pail of water. Tie a small balloon, made of goldbeater's skin, on to the tube, and having done so, press the jar down in the water, the gas will be driven in. If there is not sufficient in the jar to fill the balloon, decant another jar into it.

When the balloon is full, tie the neck just above the tube with a long piece of fine twine, and carefully cut it away from the tube. The balloon will ascend on account of the gas being lighter than air, and remain above till hauled down.

Balloons were formerly filled with hydrogen, but are now filled with coal gas, which, though heavier than hydrogen, is lighter than air.

*Exp.* 7. If a bladder be filled with the gas, and a piece of tobacco-pipe be inserted, soap bubbles may be blown with it, and will ascend ; or a piece of flexible tube, with a bit of glass tube at the end, may be connected with the generating bottle, and answer the same purpose.

57. Hydrogen is the lightest body known. Its weight, compared with air, being as 1 : 14·47. A litre of the gas at 0°C and 760$^{mm}$ pressure weighs ·0896 of a gram. A litre of air would weigh nearly 14½ times as much, or 1·3 grams nearly.

*Exp.* 8. Inflame another jar of hydrogen, and hold a dry glass over it while burning ; the glass will become covered with moisture. The hydrogen in this case combines with the oxygen of the air, and forms water.

It is best to dry the gas, however, first, to prove that water is formed by hydrogen burning in air.

*Exp.* 9. Attach a U-shaped tube, containing lumps of dried chloride of calcium, to the tube from the generating bottle, and adapt a piece of glass tube drawn out to a point to the further limb. Take care that the apparatus is sound. Ascertain that all air has been driven out of the apparatus by collecting several test-tubes of the gas and inflaming them at the lamp. When the hydrogen burns quietly down the tube, light the jet ; the gas will burn with a very pale flame, giving but little light. Hold a dry bell-jar over it, the sides will become covered with moisture, water being formed.

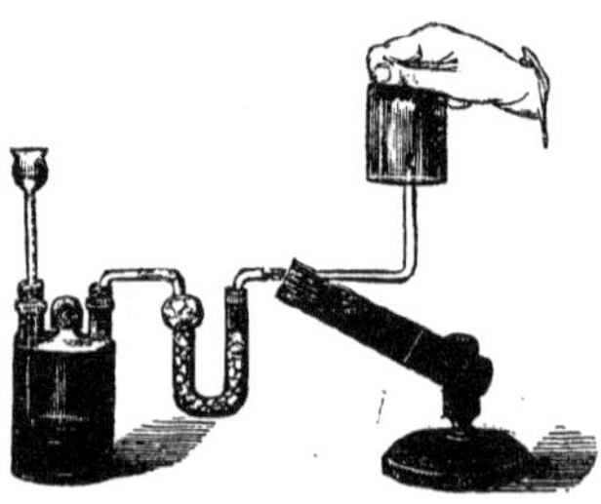

Fig. 25.

*Exp.* 10. While the jet is burning in the last experiment, hold a wide glass tube, about 18 inches long, open at both ends, over the flame, raising and lowering it at intervals ; a musical note will be produced. A rapid current of air is drawn in, which gives rise to a series of explosions which follow so quickly that a continuous sound is produced.

**58.** Hydrogen may be obtained from potassic or sodic hydrate by the action of zinc. It will be remembered that, in speaking of the action of potassium on water, it was said that *part* of the hydrogen was expelled. The other part combined with the potassium (K) and oxygen of the water to form caustic potash (OKH or KHo), thus:

$$2OH_2 + K_2 = 2OKH + H_2$$

Now the hydrogen may be replaced by zinc, but in this case one atom of Zn will perform the function of two atoms of K. And, as one atom of Zn will displace two atoms of H, two molecules of OKH must be used.

A body called potassic zinc oxide will be formed, and dissolve in the water.

$$Zn + 2OKH = ZnKo_2 + H_2$$

*Exp.* 11. Make a strong solution of potash or soda (Na Ho), and put it into a retort, with some zinc turnings. Heat it to boiling on a sand-bath; a gas will be given off which may be collected over water as usual. Some steam will also pass over, but this will condense.

On applying a light to the gas collected, it will inflame—the gas is hydrogen.

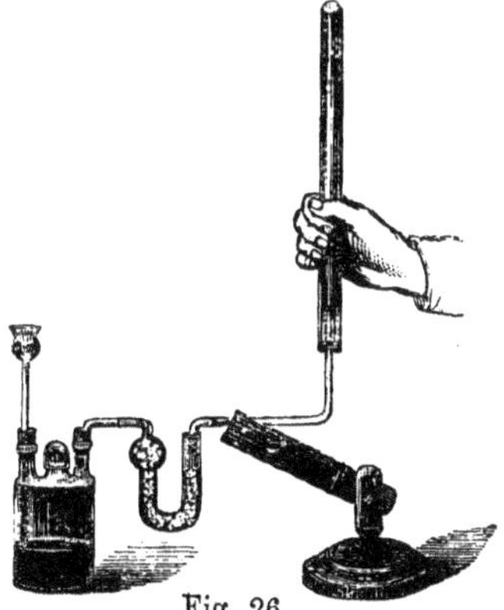

Fig. 26.

Iron, if heated to redness, decomposes water, combining with the oxygen and liberating hydrogen. A red-hot poker plunged into water does so.

*Exp.* 12. Adapt a flask, in which steam can be generated, to a piece of **hard** glass tube, in which some iron filings or borings have been previously placed. Fit a delivery-tube to this, connected with the pneumatic trough. Heat the flask, and as the steam passes over the iron filings, heat them to redness by a Here path's blow-pipe. The steam will be decomposed, triferric tetroxide, $Fe_3O_4$, being formed, and hydrogen liberated. The gas may be collected as usual.

$$Fe_3 + 4OH_2 = Fe_3O_4 + 4H_2$$

A furnace is generally used for this experiment, as in Fig. 27. Iron nails are put into a piece of gas-pipe, and the glass tubes are cemented in. Such a furnace can be easily made for a few shillings, and serves for many experiments. A mixture of coal and coke, or charcoal, may be used as fuel, and the chimney can be put into the flue of an ordinary fireplace.

Fig. 27.

The reactions which take place in *Exps.* 4 and 5, Chap. V., will now be understood:

$$2HCl + Na_2 = 2NaCl + H_2$$
$$2NH_3 + K_2 = 2KN + 3H_2$$

SUMMARY.

Hydrogen is an inflammable gas, but it extinguishes flame. It is transparent, colourless, and without taste or smell when pure. It is the lightest body known, standing in relation to air as 1 : 14·47, or ·0693 of its weight. That is, 14·47 (nearly $14\frac{1}{2}$) litres only weigh as much as one litre of air at the same temperature and pressure.

On account of its lightness, it may be poured upwards, and may be collected by upward displacement of air. Mixed with air or oxygen, it is explosive. It forms $\frac{1}{9}$ of the weight of water and $\frac{2}{3}$ of its volume. When burnt in air or oxygen, water is formed.

It is very slightly soluble in water. It is an electro-positive element, appearing at the negative pole, when compounds containing it are decomposed by the electric current. It combines with electro-negative elements, as chlorine, oxygen, nitrogen, and carbon, to form compounds from which it can be liberated by certain metals. It is also a necessary constituent of all acids, and of a large number of other compounds. It is a *monad* element, two atoms, $H_2$, being replaced by one atom of a *dyad* element, like zinc.

---

## CHAPTER IX.

### CHLORINE Cl′.

59. WE have already learned something about the properties of this gas in Chapter VI., having obtained it from hydrochloric acid by means of the electric current. It is, in fact, generally obtained from this source in the laboratory on account of its cheapness, the commercial acid being obtained as a by-product in the manufacture of carbonate of soda (washing soda) from common salt. It may indeed be obtained from common salt, NaCl.

The most simple way, however, of obtaining the gas is by heating platinic chloride, the metal platinum having but slight affinity for chlorine.

*Exp.* 1. Heat a few drops of the solution of platinic chloride in a test-tube, it will decompose, a greenish-yellow pungent-smelling gas will be evolved, which bleaches a moistened litmus paper instantly, and the metal platinum will be left in a spongy state.

$$PtCl_4 = Pt + 2Cl_2$$

Great care must be taken in experiments with chlorine, as it is a poisonous gas, and if inhaled, unless largely mixed with air, may produce serious effects. It irritates and destroys the tissues of the air-tubes of the lungs, causing violent coughing and oppression of the chest. It should not, therefore be allowed to escape into the room; indeed, all experiments with chlorine are best carried on under a hood connected with the flue, or where there is a good current of air. When not being collected, the delivery-tube should be passed into a jar containing a solution of sodic hydrate, or milk of lime, either of which absorbs the gas. It is advisable to have a cloth wetted with **ether** or **methylated spirit** laid on the table, as these liquids dissolve the gas. **Ether** may be inhaled also with advantage should the gas be inhaled by accident, but brisk exercise in the fresh air is the best restorative in such cases.

Platinic chloride is too expensive to be used where jars of the gas are required for experiments. Hydrochloric acid is the best source from which to obtain the gas in any quantity.

If HCl be heated with black oxide of manganese (manganic dioxide, $MnO_2$), the latter exchanges its oxygen for chlorine, manganic chloride ($MnCl_4$) and water being formed, the two atoms of *dyad* oxygen being replaced by four atoms of *monad* chlorine. The mangan*ic* chloride formed is a very unstable compound, splittling immediately into mangan*ous* chloride ($MnCl_2$), and setting free half the chlorine. The reactions are therefore best shown in two stages:

$$MnO_2 + 4HCl = MnCl_4 + 2OH_2; \text{ then}$$
$$MnCl_4 = MnCl_2 + Cl_2$$

*Exp.* 2. Fit up a wide-mouthed 8-ounce flask (a Florence flask will do), with a thistle funnel, and a delivery-tube bent twice at right angles. Let the longer leg of the tube pass through a cork fitted air tight into a two-necked* Woulff's bottle, and just dip into the water. Fit a delivery-tube to the other neck leading to the pneumatic trough, as in Fig. 28.

* A three-necked Woulff's bottle may be used in this case, as in making a solution of hydrochloric acid, explained in next chapter. A straight tube is fitted to the central neck, dipping into the water, so that *air* is admitted if the pressure from the generating flask diminishes, preventing a backward rush of water. On the other hand, if the pressure is too great, the water rises in the central tube. Such an arrangement is called a safety tube.

The latter must be filled with *water as hot as the hand can bear*, as chlorine is soluble in cold water to the extent of two or three times its volume.

Put about an ounce of finely-powdered manganic dioxide into the flask, pour in enough of the strongest commercial hydrochloric acid, to thoroughly wet the $MnO_2$ by shaking it. Replace the cork, and arrange the apparatus, jars being filled with water and inverted ready in the trough.

Heat the flask over a sand-bath, the chlorine will come off rapidly, any HCl that comes over without being decomposed being dissolved by the water in the Woulff's bottle. Replace the first

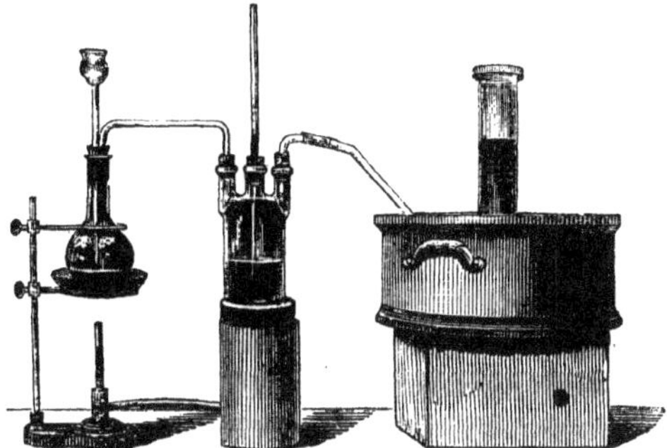

Fig. 28.

jar, when full, by another, covering the first with a greased glass plate, and removing it to the table.

*Exp.* 3. To show the solubility of chlorine, nearly fill a small-mouthed bottle with the gas, cover with the thumb and shake well; then put the mouth of the bottle under water and remove the thumb, more water will rush in, absorbing or dissolving the gas. Repeat this operation several times, and a solution of the gas will be obtained, which will do in many cases instead of the gas itself.

At 60° water absorbs its own bulk of Cl, the solution being of a pale yellow colour, with an astringent nauseous taste. The solution has the sp. gr. 1·008.

Chlorine being heavier than air, may be collected by *downward displacement*.

*Exp.* 4. Adapt a delivery-tube bent twice at right angles to the Woulff's bottle, which must be raised on the table support or on blocks, so that the outer limb of the tube shall reach to the bottom of the bottles or jars. A card with a hole cut in the middle, or a notch at the side—a little larger than the tube—being put over the jar. The gas lifts the air out on account of the difference in density. When a litmus paper is bleached at the opening, the jar may be considered full.

Of course by this mode there will be a slight admixture with air, but this does not matter for most experiments. The tube when removed from the jar should be put into one containing a solution of caustic soda, as before directed, unless others have to be filled.

The gas may be dried by passing it through calcic chloride or a bottle containing sulphuric acid and pieces of pumice stone or coke, neither of which combine with the gas.

**60.** Chlorine is an electro-negative element, appearing at the positive pole in the electric decomposition of HCl and other chlorides. It acts with great intensity upon electro-positive elements, as in the case of hydrogen (Chap. VI., *Exp.* 9), combining with and frequently liberating them from their compounds.

Antimony, arsenic, copper, etc., burn in chlorine spontaneously, forming *chlorides.*

*Exp.* 5. Shake some powdered antimony from the tip of a knife into a tall jar of chlorine. It takes fire as it descends, forming a white smoke consisting of antimonious chloride, which being volatile makes the action more violent. Finely powdered *metallic arsenic* acts in the same way. The volatilised chlorides must be avoided, as they are very poisonous. Remove the jar under the hood, or into a good draught *immediately*; or, perform the experiment there:

$$Sb_2 + 3\ Cl_2 = Sb_2Cl_6$$

*Exp.* 6. Attach a piece of Dutch metal (imitation gold leaf, consisting of copper and zinc) to a wire, and put it into a jar of Cl. It will inflame, forming cupric and zincic chlorides ($CuCl_2$ and $ZnCl_2$).

*Exp.* 7. Heat a bit of sodium in a deflagrating spoon and plunge it into a jar of the gas. The intense heat will volatilise the sodic chloride or common salt (NaCl) formed, and vivid combustion will take place. If put in without being heated, it only becomes covered with a white crust of NaCl.

*Exp.* 8. Put a little of the solution of chlorine into an evaporating dish and put a bit of sodium in it. The metal will burn, sodic chloride being formed, which will dissolve in the water and give it a *saline* taste.

**61.** Chlorine combines with chlorous or electro-negative elements, but not so energetically.

*Exp.* 9. Put a bit of phosphorous into a deflagrating spoon and plunge it into a jar of chlorine. It will burn spontaneously, but not so vividly, phosphoric chloride—$PCl_5$—being formed.

Chlorine decomposes such bodies as tallow, wax, and paraffin, which are rich in carbon and hydrogen, combining with the latter to form HCl, and liberating the carbon in a finely divided state.

*Exp.* 10. Plunge a lighted wax taper, attached to a piece of wire, passing through a card, as in Fig. 29. It will burn with a lurid flame, giving off a large quantity of smoke, white fumes of HCl being also formed in the jar.

*Exp.* 11. Steep a piece of blotting paper or rag in turpentine (not French turps). It will burst into flame instantaneously, dense volumes of smoke being evolved.

$$C_{10}H_{16} + 8\ Cl_2 = 16\ HCl + 10\ C.$$

A jet of coal gas burns with a red smoky flame in chlorine. A jet of hydrogen on the other hand burns with a feebly luminous flame, white fumes of HCl only being produced. (See Fig 30.)

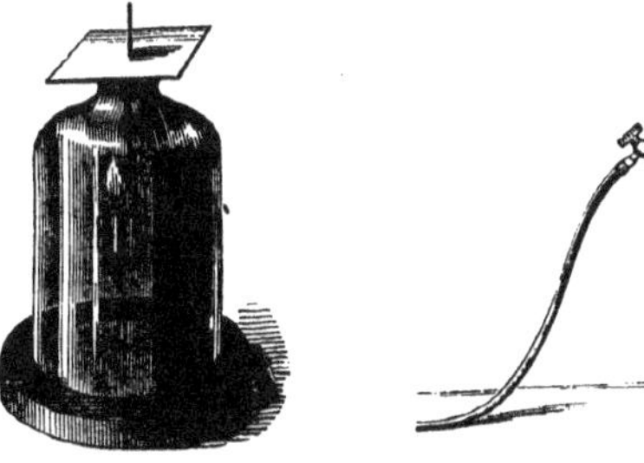

Fig. 29. Fig. 30.

**62.** Chlorine decomposes colouring matters which are obtained from animal or vegetable substances, combining with part of the hydrogen, forming colourless compounds. It is thus used for bleaching calico and other fabrics.

*Exp.* 12. Hold a bunch of violets, a bit of turkey-red cloth, or a dry litmus paper, at the mouth of a jar of Cl, the colour will be scarcely affected. Wet them and they will be instantly *bleached*. The colouring matters contain H, which is withdrawn by the chlorine, *and the oxygen of the air bleaches*. A solution of sulphate of indigo (sulphoindigotic acid) and ordinary writing ink are bleached by Cl, but printer's ink, which contains carbon, is not. The solution of the gas will do for such experiments.

The use of chlorine and *compounds* containing it, as deodorising agents, also depends on the fact that the hydrogen of decomposing animal and vegetable substances combines with it to form HCl, and the substances become oxidised by the air. All disinfectants which evolve Cl deodorise by the action of the oxygen in the air.

Slaked lime or calcic hydrate absorbs chlorine, forming the so-called "chloride of lime," or bleaching powder. For this purpose the lime is spread on shelves perforated with holes in a closed chamber, into which torrents of chlorine are sent by a pipe. The gas is absorbed to the extent of 30 or 40 per cent., but is rapidly lost on exposure to the air.

*See* Reactions, in Chap. XIII.

A weak acid—even citric or tartaric—is sufficient to liberate the chlorine from this compound. Chlorine is even evolved by the small amount of carbonic acid gas in the air, but it takes place more rapidly when mixed with water, particularly if a few drops of any acid are added.

*Exp.* 13. Put some "chloride of lime" into a beaker, and pour upon it a few drops of HCl or vinegar; Cl is evolved rapidly. Cover with a glass plate to prevent its being inhaled.

*Exp.* 14. Stamp a pattern on a piece of turkey-red cloth by dipping a bit of wood into a solution of bleaching powder ("chloride of lime"). Dip the cloth into water made *sour* by some tartaric acid. The colour will be discharged only at the stamped spots. In this way the white figures on a coloured ground are produced in coloured handkerchiefs.

It has been seen by *Exps.* 7 and 8 that common salt is a chloride of sodium. Hence the vast stores of rock-salt form the great natural storehouse of chlorine.

When sulphuric acid is poured upon common salt, both are decomposed, HCl being formed. Now, the whole of the chlorine can be obtained from the HCl at the moment of its production (when in the *nascent* state) by adding $MnO_2$ thus:

*Exp.* 15. Add together in a flask 4 parts by weight of NaCl, 1 part of $MnO_2$, 2 of sulphuric acid, and 2 of water. Heat the mixture; manganous sulphate, sodic sulphate, and water will be formed, and the whole of *the chlorine* will be evolved:—

$$2\,NaCl + 2SO_2Ho_2 + MnO_2 = SO_2Mno'' + SO_2Nao_2 + 2\,OH_2 + Cl_2$$

Some hydrochloric acid is sure to be carried over, however, unless the sulphuric acid be added through a thistle funnel. It can be got rid of by making it pass through water in a Woulff's bottle.

*Question.*—How much chlorine by weight (in grams) and by measure (in cubic centimetres) can I obtain from 1 litre of HCl, measured at 0°C and $760^{mm}$ mercurial pressure.

Here 1 litre of HCl would furnish ½ litre of Cl—

$$\therefore \frac{\cdot0896 \times 35\cdot5}{2} = 1\cdot5904 \text{ grams.}$$

And as 1 litre = a cub. dec. ½ litre = $\cdot5^{cd}$ or ·5 of 1000 = 500 cubic centimetres.

SUMMARY.

Chlorine is 35·5 times the weight of hydrogen, and $\frac{35\cdot5}{14\cdot47}$ or 2·45 times the weight of air. A litre of the gas weighs $\cdot0896 \times 35\cdot5 = 3\cdot1808$ grams.

Cold water dissolves its own volume of the gas when agitated. The gas is of a yellowish-green colour, and is very pungent and poisonous. It forms salts with the metals called chlorides. It is a bleaching and disinfecting agent on account of its affinity for hydrogen. A solution of the gas acts like the gas itself.

---

## CHAPTER X.

### Hydrochloric Acid HCl.

63. This is a gas; its solution is known by the same name; it is also called muriatic acid or spirits of salt, being used in the laboratory and the arts for a variety of purposes.

It has already been prepared synthetically by exploding a mixture of equal volumes of hydrogen and chlorine.

$$H_2 + Cl_2 = 2HCl.$$

The same reaction takes place when a jet of H is burned in Cl.

Combination of the gases H and Cl also takes place slowly in diffused daylight, and immediately with explosion in strong or direct sunlight, or in a light rich in chemical rays.

*Exp.* 1. Fill a jar, half with Cl and half with H. Cover it with a greased glass plate, and put it in the *dark*, no action takes place. Remove the jar into the daylight—not in the blaze of the sun; slow combination takes place, HCl being formed, which will redden a blue litmus paper, and not bleach it unless the Cl was in excess. White fumes will appear in the jar if it or the gas be moist.

If a small glass bulb be filled with the mixed gases, it may be exploded in the following manner:—For safety put it into a cigar-box turned on its side. Into a cylindrical jar filled with nitric oxide (for preparation *see Exp.* 11, page 121), pour in a few drops of bisulphide of carbon, or put a small bulb containing that liquid into the jar; shake up and break the bulb, so that the $CS_2$ may be spread over the sides of the jar. Place the jar before the open box containing the bulb of mixed H and Cl. *Stand behind the box,* and removing the glass plate that covers the jar, apply a lighted taper; the mixture will burn with a blue flame rich in chemical rays, which will cause the H and Cl in the bulb to combine with a loud crack, breaking the glass, white fumes being produced from the hydrochloric acid gas dissolving in the moisture of the air.

**64. Hydrochloric acid** is prepared by gently heating sodic chloride (common salt) with sulphuric acid, diluted previously with a small quantity of water. The sodic chloride is decomposed and also the acid, one half the hydroxyl of the latter being replaced by sodoxyl, forming hydric sodic sulphate, which dissolves; the other half of the hydrogen combining with the chlorine of the common salt.

$$NaCl + SO_2Ho_2 = SO_2HoNao + HCl.$$

*Exp.* 2. Into an 8-oz. flask fitted with a funnel and delivery-tube (or a Florence flask), as in making chlorine, Fig. 28, put three parts by weight of fused sodic chloride and five of oil of vitriol, mixed with an equal quantity of water. Heat gently on a sand-bath; a colourless gas will come off, which being rather heavier than air, may be collected by displacement as in the case of chlorine.

It cannot be collected over water on account of its great solubility, but it may be over mercury, though it is not advisable, as the gas, which is very irritating to the eyes, is apt to escape, and the experiment is somewhat too difficult for a young student to manage.

*Exp.* 3. Collect two small cylinders of the gas, and into one jar plunge a lighted taper; it will be extinguished. A blue litmus paper will be turned red, showing the gas to have acid properties.

*Exp.* 4. Invert the second jar under water, which may contain a small quantity of infusion of litmus. Remove the greased plate, and the water will rush violently up the jar, dissolving the gas completely, and turning the litmus red. To perform this experiment well, the gas should be dried by passing through a dessicating bottle containing sulphuric acid, or through a tube containing chloride of calcium (*see* Fig. 15, p. 46).

Fig. 31.

Should any be collected over mercury, it must first be dried for the latter experiment, and then a piece of ice may be inserted under the jar, water even in the solid form aborbing the gas.

The following experiment shows the solubility of the gas in a striking manner:—

*Exp.* 5. Collect a jar of dry HCl over mercury, and then pour some water into the vessel. Raise the jar out of the mercury into the water—the latter will rush up and fill the jar (Fig. 31).

A glass tube, with a stout bulb at the end, may be used for showing the solubility of the gas. This being filled with the dried gas, should be inserted in a little cup of mercury, which should then be put into a vessel of water; the tube being raised above the mercury, the water will rush up with great violence. Water dissolves 480 times its volume of the gas, the solution being increased by about $\frac{1}{3}$ in bulk, with a specific gravity of 1·2 compared with water.

**65. Hydrochloric acid converts certain metals, metallic oxides, and hydrates, into salts called chlorides.** Indeed, it may be considered as a salt of hydrogen, and is hence frequently called **hydric chloride.** The salts derived from it by the substitution of a metal for hydrogen being called **haloid** salts, to distinguish them from those containing oxygen.

Thus, Zn, Fe, Mg, and Al, are dissolved by Hcl, liberating hydrogen and forming chlorides. And here the knowledge of **atomicity** becomes valuable, enabling the student to know how many atoms of hydrogen can be replaced by the metal, and hence how many molecules of the acid are required. The dash on the left of the symbol for aluminium shows the latent bond.

$$Mg_2'' + 4HCl = 2MgCl_2 + 2H_2$$

$$'Al_2''' + 6HCl = \left\{ \begin{matrix} 'Al'''Cl_3 \\ 'Al'''Cl_3 \end{matrix} \right. + 3H_2$$

The dried gas passed over the metals potassium and sodium in Chap. V. acted in the same manner (*see* also Chaps. VII. and VIII.).

Some metals are but slightly attacked by this acid when cold, as Pb, Ag, Cu, Bi, Sb, and the metal tin only with difficulty even in boiling acid.

*Exp.* 6. Put a little cupric oxide (black oxide of copper) into a test-tube, and pour upon it some strong HCl. Gently heat it, a green solution of cupric chloride will be the result. The dried gas may be passed over it in a hard glass tube heated by a Bunsen flame, with the same result.

$$CuO + 2HCl = CuCl_2 + OH_2$$

The oxides of zinc and mercury act in a similar manner.

*Exp.* 7. Make a solution of sodic or potassic hydrate in a beaker, and add a solution of HCl drop by drop to it till it neither turns a reddened litmus paper blue, nor a blue one red. The solution will be neutral and have a saline taste. If some of it be evaporated slowly, crystals of sodic or potassic chloride will be formed.

$$NaHo + HCl = NaCl + OH_2$$

Baric and calcic chlorides may be prepared from their corresponding hydrates in the same manner; water being a secondary product in all cases.

$$CaHo_2 + 2HCl = CaCl_2 + 2OH_2$$

All chlorides but those of Pb, Ag, and Hg, are soluble in water, the former of these being soluble in boiling water, and AgCl in ammonia.

The test for chlorides in solution is argentic nitrate $No_2Ago$ (nitrate of silver). The strongest solution contains about 40 per cent. of acid. If this be heated, the gas is liberated till the density becomes 1·1, when both acid and water distil over at 106°C unchanged. A weaker acid loses water till its density is 1·1 (containing only about 20 per cent. of gas), and then at 106°C behaves as above. The strength of a solution can be tested by the hydrometer.

To prepare a solution of the gas for laboratory purposes, it is passed through a series of three or four Woulff's bottles half-filled with distilled water. The delivery tubes must only just touch the surface of the water, as the gas is absorbed with great avidity, and the bottles must be furnished with safety tubes. (*See* preparation of Chlorine.)

The first bottle retains any sodic chloride or sulphuric acid that might have been carried over mechanically, and the water in the other bottles become successively saturated with the gas.

Cast-iron cylinders lined with fire-clay heated in a furnace, and earthenware bottles, are used in its preparation on the large scale. The crude acid obtained in this way is often *yellow* from the iron of the retorts in the form of *ferric chloride;* arsenic, as arsenious chloride, is also present sometimes from the impure sulphuric acid used. If $SO_2Ho_2$ be present, the addition of a solution of *baric chloride* produces a white turbidity (a precipitate = pp.). Arsenic gives a yellow colouration or pp. with sulphuretted hydrogen; and the addition of a solution of ferrocyanide of potassium gives a pp. of prussian blue if *ferric chloride* be present. When pure, the solution should be perfectly colourless, and give no pp. with the above reagents.

Hydrochloric acid is obtained as a by-product in the

manufacture of carbonate of soda from common salt. The sodic chloride is put into a reverberatory furnace, and oil of vitriol is poured upon it through a funnel opening in the roof. The flames play over a bridge upon the materials, and in consequence of the great heat obtained, the whole of the hydrogen of the acid combines with the chlorine, or the whole of the hydroxyl is exchanged for sodoxyl, forming the neutral sulphate of soda. In consequence of this, one half the quantity of acid only has to be used, as in the following equation:

$$2\,NaCl + SO_2Ho_2 = SO_2Nao_2 + 2HCl$$

Formerly, torrents of the gas escaped as a waste product from the chimneys of the manufactories in the North of England, poisoning the neighbourhoods where they were situated.

It is now conveyed into towers filled with coke, down which a stream of water constantly trickles and dissolves the gas. The towers are connected with the chimney to produce a draught. A solution is obtained in this way, which is used in the preparation of chlorine in the manufacture of bleaching powder ("chloride of lime"). *See* Chap. XIII.

---

## CHAPTER XI.

### OXYGEN O''.

**66.** OXYGEN is a **dyad** element. It is very abundant in nature, forming $\frac{8}{9}$ of the weight of water. It forms also $\frac{1}{5}$ of the volume of the atmosphere, the remainder being principally nitrogen, with which it forms a mechanical mixture. It forms also about $\frac{1}{3}$ of the solid crust of the globe, being found as a constituent of such minerals as lime, quartz, clay, as well as in combination with various metals as oxides.

It has already been obtained from water by the action of the electric current (*see* Chap. VI.). It can also be obtained by chemical means as stated in *Exp.* 10 of the present chapter.

It may be abstracted from the air by keeping mercury near its boiling point for several hours. The red powder thus obtained when heated to about 400° C gives up its oxygen again, so that this is a means of obtaining oxygen from air. A molecule of HgO (mercuric oxide), or 216 parts by weight, yielding 200 parts by weight of mercury, and 16 of oxygen, which are the relative weights of the atoms of these elements. Or, to express the decomposition by a molecular equation, so that a molecule of oxygen should be liberated,

$$2\ HgO = Hg_2 + O_2$$

*Exp.* 1. Heat a small quantity of mercuric oxide in a test-tube or bit of hard glass in the Bunsen flame. Adapt a delivery-tube so that the gas may be collected at the pneumatic trough. A mirror will appear round the side of the tube, which will be found to consist of globules of metallic mercury, and a colourless gas will come off which may be collected in a test-tube as usual.

Plunge a glowing splinter into the gas, it will be rekindled. The gas is oxygen.

Argentic oxide may be used, but it requires greater heat to decompose it—that of a Herepath's blow-pipe for instance. Oxygen will be given off and a button of silver will be left.

$$2\ OAg_2 = 2\ Ag_2 + O_2$$

Both the latter methods are, however, expensive, and therefore other materials must be resorted to where any quantity of the gas is required.

The mineral pyrolusite, black or dioxide of manganese, used in obtaining chlorine from HCl, Chap. IX., gives up one-third of its oxygen when heated, but the experiment requires to be done in an iron bottle. Another compound of manganese and oxygen being produced, similar to that of iron, by the passage of steam over it as in *Exp.* 12, Chap. VIII. It is called the trimanganic tetroxide.

$$3\ MnO_2 = Mn_3\ O_4 + O_2$$

If it be heated with sulphuric acid, it gives up one-half its oxygen, and sulphate of manganese is formed.

$$MnO_2 + SO_2Ho_2 = SO_2Mno'' + O_2 + OH_2$$

*Exp.* 2. Put a small quantity of dry powdered manganic dioxide into a test-tube, and pour sufficient strong sulphuric acid on it to thoroughly wet it. *Heat cautiously.* Oxygen gas will come off, which may be tested by a glowing splint.

The salt called potassic chlorate (chlorate of potash) is the most convenient source of oxygen in the laboratory. Some of it should be first put into an evaporating dish and heated. The salt will melt, and the water held in the crystals will be given up. It hardens on cooling, and should then be pulverised in a mortar.

It may be either used alone, or it may be mixed with about one-third of its weight of manganic or cupric oxides (powdered glass or sand will do); the gas comes off at a lower temperature, but is not quite so pure. The manganic dioxide appears to be unaltered at the end of the experiment. It seems to attract more oxygen from the chlorate, with which it cannot however combine, and thus assists the repulsive action of heat.

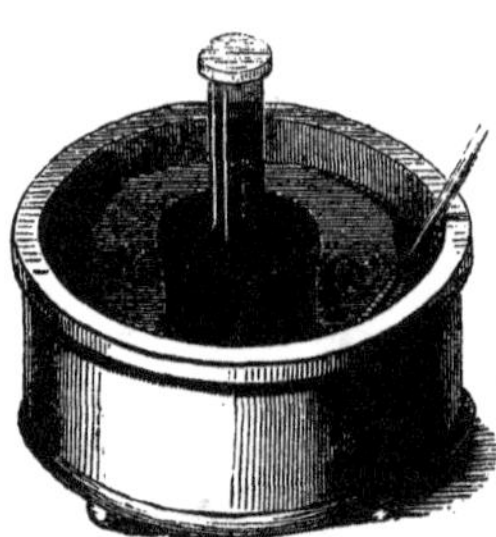

Fig. 31-*a*.

The three atoms of oyygen in the potassic chlorate are all given off in either case, as one of them cannot exist alone, hence two molecules are used in the graphic and symbolic equations below.

$$\text{Graphic} \left\{ \begin{array}{l} K - O - O - O - Cl \\ K - O - O - O - Cl \end{array} \right\}$$

$$\textit{Heated} = \left\{ \begin{array}{l} K - Cl \\ K - Cl \end{array} \right\} + \left\{ \begin{array}{l} O - O \\ O - O \\ O - O \end{array} \right\}$$

$$\text{Symbolic (better method) } 2 \left\{ \begin{array}{l} OCl \\ O \\ OK \end{array} \right. = 2\,KCl + 3\,O_2$$

Potassic perchlorate is, however, first formed (*see* Chap. XIII.), and this again is decomposed into potassic chloride and oxygen.

$$(\text{Potassic Chlorate})\ 2\ ClO_2Ko = ClO_3Ko + KCl + O_2$$
$$\text{Then } ClO_3Ko = KCl + 2\,O_2$$

*Exp.* 3. Put some of the mixture (sold as oxygen mixture), which has been previously dried, into a dry Florence flask fitted with a delivery-tube for collecting the gas in jars over water in the pneumatic trough (Fig. 31-*a*).

The residue left in the flask should be allowed to cool, and afterwards water should be poured in and agitated. The potassic chloride will dissolve, and the whole should be thrown on to a large filter. The filtrate will contain the KCl, which is of no service, but the black residue on the filter $MnO_2$, may be well dried and used again.

For storing large quantities of such gases as oxygen and hydrogen a gas-holder is very useful. One called Pepy's is shown in Fig. 32.

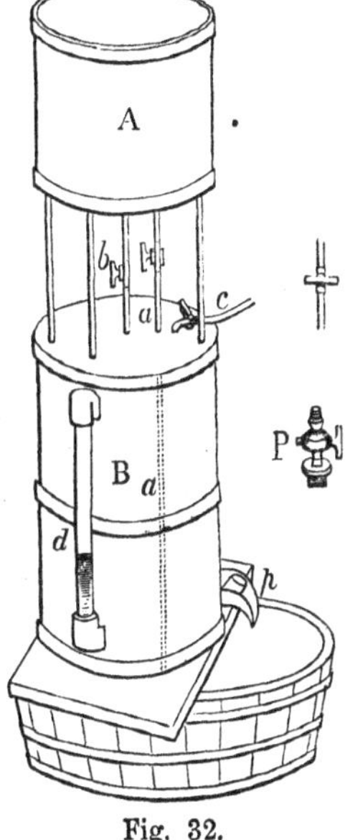

Fig. 32.

It consists of two cylinders, A and B, made of zinc, connected with each other by two tubes. One of these, *a*, passes nearly to the bottom of the lower cylinder, the other *b*, just through its upper part. There is also a third bent tube *c* passing out of the top of the lower cylinder, to which a flexible tube can be attached. These are all furnished with stopcocks. A tube passes out obliquely from the bottom of the lower cylinder, which can be closed by a screw plug *p*. To use the gas-holder, it is placed over a sink, or tub, as in Fig. 32. The plug is screwed in, the taps of *a*, *b*, and *c* are opened, water is poured into the upper cylinder or funnel A, and runs down *a* and *b*, filling the lower cylinder B, the air coming out at *c*, and the height of the water being shown by the gauge *d*.

When the lower cylinder is filled with water, the stopcocks *a*, *b*, and *c* are closed, and the screw plug taken out, the water being kept in the cylinder by the pressure of the atmosphere. The gas delivery-tube is inserted at *p*, and the gas rises to the top of B, driving out the water at *p* into the tub.

When full of gas, the plug is screwed in, and water poured into the funnel A. The stopcock *b* is opened to allow water to flow into B to press out the gas. Jars of gas may be filled at A as in a pneumatic trough by opening the stopcock *b*, or a flexible tube may be attached to *c*. Water must of course be continually poured into A to keep up the pressure in B, and the latter will be full of water as shown by *d* when all the gas has been used.

Having collected several jars of the gas either from the gas-holder or at the pneumatic trough in the usual way, try the following experiments:—

Fig. 33.

*Exp.* 4. Plunge a piece of small wax taper stuck on the end of a wire passing through a card into a jar of oxygen. It will burn very fiercely till all is consumed. Cover the jar with a glass plate for further examination.

*Exp.* 5. Wind a piece of wire passed through a card round a piece of charcoal, taken from the bark. Heat it to redness in the Bunsen flame, and plunge it into a wide gas jar; it will burn with beautiful scintillations, a colourless gas being the result. Cover this and put it aside for further examination also.

*Exp.* 6. Put a small piece of sulphur into a deflagrating spoon run through a bit of card or attached to a brass flange. Melt the sulphur and it will inflame. Plunge it into another bell jar of oxygen. It will burn with a beautiful blue flame, producing a colourless gas with a pungent odour. Cover and put aside.

*Exp.* 7. Inflame a small chip of phosphorus in a deflagrating ladle, and plunge into a bell-jar of oxygen. It will burn brilliantly, a dense white vapour being formed, which itself takes fire, and makes the light of dazzling brilliancy. Cover and put aside.

*Exp.* 8. Make a spiral of very fine iron wire by coiling it round a pencil. Pass it through a card. Put a bit of German tinder on the end of it and light it, or dip it into melted sulphur. Plunge it into a bottle of the gas. Beautiful scintillations will be thrown off, consisting of the iron combined with oxygen, which often fuse into and crack the glass, hence a little water should be left in the bottom. A piece of watch-spring, first heated to redness to soften it, and then plunged in quickly whilst red-hot, may be substituted for the iron wire.

*Exp.* 9. Melt a bit of sodium in the deflagrating ladle, which should first be cleaned and dried. It will take fire, and when plunged into oxygen, will burn more brilliantly, with an intensely yellow flame.

Turning to the results of these experiments, we find:

*Exps.* 4 and 5. A colourless gas in both cases. Test with a lighted paper, the gas will extinguish it. Pour in some *lime-water* to each, it will be turned milky. The result of burning the taper or charcoal in oxygen was to produce a gas called carbonic acid gas, $CO_2$, which combines with the lime, and forms chalk or carbonate of lime (calcic carbonate, $COCao''$).

*Exp.* 6. Put into this jar a moistened blue litmus paper, it will be turned red, the gas formed by the combination of sulphur and oxygen having acid properties. It is called sulphurous anhydride, $SO_2$.

*Exp.* 7. The white vapour will be found to have dissolved in the water on the sides of the jar. A blue litmus paper in this instance too will be reddened. The white substance formed was phosphoric anhydride, $P_2O_5$, which, dissolved in water, formed phosphoric acid.

$$P_2O_5 + 3OH_2 = 3POHo_3$$

In *Exps.* 8 and 9, the metals, iron and sodium, combined with oxygen, forming two of a class of solid bodies without acid properties, called oxides. In the first case, the triferric tetroxide (magnetic oxide of iron), $Fe_3O_4$, and in the second, sodic oxide, $Na_2O$, were produced.

**67.** Many other metals, when exposed to air or oxygen, are converted into *oxides*, or, in other words, *rust*. Zinc and iron do so at the ordinary temperature, forming $ZnO$ and $Fe_2O_3$ (sesquioxide of iron); copper does so at a red heat; whilst gold and platinum do not oxidise at the most intense heat.

A stream of oxygen from the gas-holder, or a bladder filled with the gas, in which a tobacco pipe has been inserted, made to play upon a piece of watch-spring whilst ignited at the Bunsen flame, gives rise to a brilliant combustion. Such effects, however, are usually obtained by the oxyhydrogen blow-pipe, and are more fitted for the lecture table. A stream of oxygen is made to mix beforehand with a stream of hydrogen or coal gas, and the two are burned at a jet, a very small but very hot flame being the result, in which case, iron, and even platinum, can be melted, the former with beautiful scintillations, producing $Fe_3O_4$; the latter, however, being unchanged on cooling. The mixed gases made to impinge on a cylinder of lime give rise to the intense lime-light.

**68.** Oxygen may be obtained from water, through the agency of chlorine, on account of the great affinity of the hydrogen for chlorine. When steam and chlorine are passed through a *porcelain* tube heated to redness in a furnace, HCl is formed, and oxygen is set free, as seen in the annexed figure.

Fig. 34.

*Exp.* 10. Arrange a flask for generating chlorine, as in Chap. IX., *Exp.* 2. Let the delivery-tube from this pass into the water of a second flask, in which steam is being generated (Fig. 34). The moist chlorine should pass through a Woulff's bottle, where any superfluous water may condense, to prevent the cracking of the porcelain tube. The hydrogen will combine with the chlorine in the red-hot tube, and the hydrochloric acid formed will be absorbed by the water in the pneumatic trough, and the oxygen may be collected in a receiver.

Oxygen is taken away, or, in other words, **de-oxidation**

or **reduction** takes place, when a stream of hydrogen from a generating flask or a gas-holder is passed over heated cupric oxide. The experiment may be performed by putting some of the dried CuO in a hard glass tube, and heating with the Bunsen flame or Herepath's blow-pipe. The test for oxygen is nitric oxide.

*Exp.* 11. Pass bubbles of nitric oxide (prepared in the way described in *Exp.* 12, Chap. XVII.) into oxygen. Red fumes are produced.

**69. Ozone** is a modification of oxygen, first discovered by Schönbein, and so named by him from its odour. The peculiar smell of the air near an electric machine in action, or during a thunderstorm, is due to a part of the oxygen being converted into ozone. It is condensed oxygen, three atoms of oxygen being squeezed into the space of two to form the molecule.

*Exp.* 12. Put a stick of phosphorus into a bottle containing a little water, and let it remain for some hours. Ozone will be produced. Pass a strip of paper through some starch-paste, and then through a solution of potassic iodide. Dip it into the jar, the iodine will be set free, and colour the starch-paper blue.

$$6KI + O_3 + 3OH_2 = 6OKH + 3I_2$$

*Exp.* 13. Pour some strong sulphuric acid into a beaker containing a solution of permanganate of potash. Oxygen, in the state of ozone, will be immediately set free. Apply the starch-paper test, as before.

In the electric decomposition of water, the oxygen is in the state of ozone, and it may be readily obtained by the electrolysis of dilute sulphuric acid and chromic acid. It has also been obtained, by Siemens, by an induction arrangement.

Ozone is a powerful oxidising agent, silver and mercury are corroded by it at the ordinary temperature, and so are substances like cork and caoutchouc. It is insoluble in water, but it is completely absorbed by oil of turpentine. It is decomposed slowly at 100°C, and rapidly at 290°C, returning to its former volume.

The starch-test named above is not, however, very reliable, as strong acids act in the same manner :—

$$4KI + 4HCl = 4HI + 4KCl$$
$$\text{and } 4HI + O_2 = 2OH_2 + 2I_2$$

Hydrochloric and sulphuric acids are generally contained in the air of towns; country air, and especially that near the sea, has, however, the same effect, liberating iodine from the iodized starch-paper, and such air cannot be supposed to contain these gases.

### SUMMARY.

Oxygen is colourless, transparent, inodorous, and tasteless. It is rather more soluble in water than hydrogen, at 0°C it dissolves about 4 per cent. of the gas.

It is from the oxygen dissolved in water that fishes obtain the means of aërating their blood. It enters into combination with a large number of bodies producing oxidation, in the case of the metals forming a class of bodies called oxides. It is the source of combustion in air, of which it forms $\frac{1}{5}$ of the volume. It is rather heavier than air, being as 1·1056 to 1, and 16 times the weight of H—a litre weighing ·0896 × 16, or 1·4436 grams. It is usually obtained from manganic oxide or chlorate of potash. The former gives up one-third of its oxygen by heat alone, and one-half when heated with sulphuric acid. Potassic chlorate gives up all its oxygen when heated.

It is well for the student to practise himself with calculations on the quantity of materials required to produce a certain volume of gas, to find the weight of such volumes, and of the secondary products. Examples of these are given below, and a few others will be given hereafter.

1. What quantity of oxygen by weight, and also by volume, can be obtained by the decomposition of 100 grams of potassic chlorate, $ClO_2Ko$ ($KClO_3$)?

Here a molecule of $KClO_3$ would weigh 39+35·5+48 (16 × 3) =122·5, which would furnish 48 parts by weight of oxygen.

Hence, to find how much 100 grams would produce:

$$\text{As } 122{\cdot}5 : 100 :: 48 ;$$

$$\text{or, } \frac{100 \times 48}{122{\cdot}5} = 4800 \div 122{\cdot}5 = \mathbf{39{\cdot}18\ grams.}$$

Now, a litre of oxygen weighs ·0896 gram × 16 or 1·4336 grams Hence, if we divide the weight of oxygen produced by a given weight of $KClO_3$, by the weight of a litre, it will give the volume in litres.

*i.e.*, 39·18 ÷ 1·4336 = **27·32 litres.**

2. How much zincic oxide, ZnO, can be obtained by the oxidation of 100 grams of zinc?

Here the atomic weight of zinc being 65, and oxygen 16, the molecule of ZnO would weigh 65+16, or 81.

Then, as 65 : 100 : : 81 ;

or, $\frac{81 \times 100}{\cdot 65}$ = 8100 ÷ 65 = **124·6** grams.

3. A stream of dry hydrogen being passed over 100 grams of cupric oxide (CuO) heated to redness in a hard glass tube, how much metallic copper would be left, and what weight of water would be formed?

$$CuO + H_2 = OH_2 + Cu.$$

That is, a molecule of CuO and a molecule of hydrogen would produce a molecule of water and leave an atom of Cu—*i.e.*, 63·5+16, or 79·5 CuO and 2H = 16 + 2 (or 18) of water + 63·5 Cu.

Hence as 79·5 : 100 : : 63·5 : 79·87 copper.
And as 79·5 : 100 : : 18 : 22·64 water.

4. What weight and volume of carbonic acid gas would be produced by burning 5 grams of carbon in oxygen gas?

In this case, one atom of C weighing 12 combines with two atoms of O weighing 32, producing a molecule of $CO_2$ weighing 44. This molecule is two volumes. Hence, 1 volume=22; and in grams, 1 litre would weigh ·0896 × 22, or 1·9712 grams. Hence to find the weight.

As 12 : 5 : : 44, or $\frac{5}{12}$ of 44 = 18·3 grams.

And, as above, $\frac{18 \cdot 3}{1 \cdot 9712}$ = 9·3 litres of $CO_2$.

---

## CHAPTER XII.

### Oxygen Compounds with Hydrogen, Water, and Hydroxyl.

**70.** Oxygen forms two compounds with hydrogen; **water**, and **hydroxyl**, their graphic and symbolic formulæ being given below:

Water, H—O—H or $OH_2$

Hydroxyl, H—O—O—H or $\left\{ \begin{matrix} OH \\ OH \end{matrix} \right.$ or $Ho_2$ or $H_2O_2$

**71. Water** is very widely diffused. It always exists in the air in an invisible state, and gives the blue appearance to the sky, and becomes visible in the form of clouds. It forms a constituent of all animal and vegetable substances, and of many minerals. It is also formed in a variety of chemical processes as a secondary product.

The composition of water was first determined by Cavendish, who exploded a mixture of two volumes of hydrogen with one of oxygen in a stout glass tube by means of the electric spark, watery vapour being the result, which condensed on the sides of the tube. A more exact method, by which the resulting volume can be ascertained, is to enclose a stout tube containing the mixture in a jacket or outer tube in which the vapour of boiling analine is conveyed. This prevents the resulting water gas from condensing, and it is thus found to contract to two-thirds the original volume.

| | | |
|---|---|---|
| 1 volume of oxygen | or by weight | O = 16 |
| 2 volumes of hydrogen | ,, | H = 2 |
| = 2 volumes of water gas | | OH = 18 |

*Exp.* 1. Fill a eudiometer with mercury, and let it rest on a caoutchouc pad in a trough of mercury, supporting it by the wooden vice, Fig. 7, or a clip attached to the retort stand. Introduce 180 c.c. of hydrogen, and afterwards 90 c.c. of oxygen (never more than about half fill the eudiometer). Grasp it firmly by the hand, pressing it down upon the india-rubber pad, and pass a spark through the mixture by means of an induction coil or Leyden jar. A flash will pass through the mixture, and sudden expansion will first take place from the heat produced by the combination. When cool, the gas will be found to occupy 180 c.c., showing that 3 vols. have contracted to 2. The gas will, however, soon condense on the sides of the tube, and the mercury will rise.

In the bent eudiometer the thumb is held at the open end, and acts as a buffer.

The composition of water may be determined by taking any quantity of each gas, as in the following example: Introduced 8·5 c.c. of oxygen, then 19·9 c.c., or rather more than double the volume of hydrogen. This gave

28·4 c.c. Exploded and allowed to condense, 2·9 c.c. remained, which proved to be hydrogen. Then 28·4 − 2·9 = 25·5, of which $\frac{1}{3}$ was O and $\frac{2}{3}$ H. Hence, 17 c.c.

Fig. 35.

of hydrogen, combined with 8·5 c.c. of oxygen, or in the proportion of 2 to 1.

From its composition by weight the percentage composition of water can easily be found, thus :

As 18 : 100 : : 2 : 11·11 of hydrogen.
And 18 : 100 : : 16 : 88·99 of oxygen.

Its decomposition by the electric current has been explained in Chap. VI., and its composition has been determined also by the experiments in Chaps. V. and VI.

Water is formed when a jet of hydrogen is burnt in oxygen or in air. It is also formed with other products when tallow, wax, gas, etc., are burnt, and also in the combustion of coal.

Fig. 36.

*Exp.* 2. Bring the flame of a spirit lamp under a funnel which leads into a glass tube surrounded by another tube containing water (a water jacket),

Liebig's condenser may be used for the purpose, having a wide-mouthed test-tube attached to the other end containing a little *lime-water;* water will condense in the test-tube, and $CO_2$ will also be formed, which upon agitation will turn the lime-water milky.

A beaker held over a flame will show the same result, but not quite so well.

The action of water upon metallic oxides is sometimes very energetic. This is well seen in the slaking of lime.

*Exp.* 3. Get a bit of quick lime, put it on a plate and pour a little water on it, it hisses and steam issues from it, the mass cracking from the expansion produced by the heat of the chemical action. A thermometer inserted into the mass will show the temperature of boiling water.

$$CaO + OH_2 = CaHo_2$$

**72. Calcic hydrate** is produced. Some of it may be put into a large bottle, which may be filled with cold distilled water and thoroughly shaken. A part of the lime will dissolve, forming *lime-water*, the rest will fall to the bottom as milk or cream of lime, which may be dissolved hereafter. 200 parts of water dissolve only one part of lime. The solution should be kept still, and will be perfectly clear. Should it go turbid, it must be filtered before use.

Water transforms anhydrides into acids, as

$$N_2O_5 + OH_2 = 2\ NO_2Ho$$
$$SO_3 + OH_2 = SO_2Ho_2$$
$$P_2O_5 + 3\ OH_2 = 2\ POHo_3$$

Water also forms an important constituent of the molecules of certain crystals, which fall to powder when this water is driven off by heat. Such are the alums, which crystallise with $24OH_2$, and sulphate and nitrate of copper, which crystallises the former with $5OH_2$, and the latter with $3OH_2$.

Some salts lose water on exposure to the air, or *effloresce*, as carbonate of soda. Others, again, absorb moisture from the air, and become wet—*deliquesce* (go to water)—as carbonate of potash, calcic, sodic, and magnesic chlorides.

Water has great solvent powers. The solution of solids is usually increased by heat, but a solution saturated when hot generally gives up the solid in a crystalline form on cooling.

*Exp.* 4. Dissolve some acetate of lead in a test-tube, add a few drops of HCl; a white pp. will fall of chloride of lead. Add water, and boil, the pp. dissolves, but upon cooling, by plunging the test-tube into cold water, the plumbic chloride will separate again in a crystalline state.

Water on solidifying gives out the same quantity of heat that ice absorbs on melting. A parallel case in the crystallisation of sulphate of soda may be here introduced.

*Exp.* 5. Dissolve in a 4-oz. flask as much sulphate of soda $(SO_2Nao_2, 10\,OH_2)$ as the water will take up when boiling. If the solution be now corked and kept perfectly still, it will often remain for days without crystallising. But if a crystal or the point of a glass rod be introduced, or the liquid be even shaken, crystals will shoot through the liquid, and the mass will become solid—heat being given out.

On the other hand, by dissolving this salt in cold water, heat is abstracted. On this principle the formation of freezing mixtures depends. The following are a few examples of such :—

| Without Ice. | Pts. by measure | Cold Produced. | With Ice. | Pts. by wt. | Cold Produced. |
|---|---|---|---|---|---|
| Nitrate of ammonia<br>Water, . . . | 1<br>1 | 10° to 15°C | Snow or crushed ice,<br>Common salt, . | 2<br>1 | 21°C |
| Sulphate of soda, .<br>Dilute nitric acid, . | 3<br>2 | 10° to 19°C | Snow, . . .<br>Salt, . . .<br>Nitrate of ammonia | 12<br>5<br>5 | 31°C |
| Phosphate soda, .<br>Nitrate ammonia. .<br>Dilute nitric acid, . | 5<br>3<br>4 | 10° to 37°C | Snow, . . .<br>Chloride calcium, . | 2<br>3 | 0° to 45°C |

On account of its solvent powers, water is never found perfectly pure in nature. Thus, although water evaporated from seas and rivers may be so, as it descends in rain it dissolves gases and solids in the air, and when it percolates through the soil, solids of various kinds are dissolved or held in suspension. Hence the water of our springs contains from 5 to 20 grains of solid matter in the gallon. This gives rise to the term *mineral waters.* Some of the calcareous springs near London contain as much as 18 or 20 grains of chalk to the gallon, or ·028 of a gram to a litre. This chalk is held in solution by the carbonic acid gas which the water contains, and causes such water to be *hard,* making it difficult to form a lather with soap. The hardness, however, of such water is only temporary, as it may be removed by boiling, by long exposure to the air, or by the addition of carbonate of soda. Other waters which contain sulphates are permanently hard, and this hardness cannot be removed.

**73.** We have waters also holding salts of iron in solution, called *chalybeate,* and many others.

Besides the substances dissolved in water, there are two other kinds of impurities.

1. *Mineral substances* held in suspension, which can be removed by filtration.

2. *Organic substances* either dissolved or held in suspension, the former not being capable of removal.

Perfectly pure water corrodes lead, but water containing carbonic acid acts as a preservative. Water has several remarkable properties, which can only be slightly touched upon here.

Its point of greatest density is about 4°C, or rather above the freezing point. As it gets colder it expands, so that ice occupies more space than the water did before freezing. If this were not so, our lakes and ponds would become solid masses of ice in winter, and all living things in them would die.

*Exp.* 6. Put some water in a large test-tube surrounded by ice in a basin, and put the bulb of a thermometer into it. The water will fall gradually in the tube till just above the freezing point, when it will again expand and rise.

Water has a very high *specific heat,* that is, it requires a large amount of heat to raise a certain quantity through a certain number of degrees.

*Exp.* 7 (*a*). Mix equal weights of hot and cold water in two beakers, marking the temp. of each. The mixture will be the *mean* of the two.

(*b*). Again, mix equal weights of cold water and hot mercury, marking the temp. of each as before. The water will require so much heat to raise its temp. that the result will be less than the *mean*.

(*c*). Thirdly, mix equal weights of cold mercury and hot water, the temperature will be greater than the *mean* of the two.

The *latent heat* of water is also very high *i.e.*, 79° C. That is, the amount of heat required to melt a certain weight of ice, would be sufficient to raise the temperature of the same weight of water 79° C.

*Exp.* 8. Put some crushed ice or snow into a beaker, and insert the bulb of a thermometer. Heat the beaker over a sand-bath or wire gauze, wiping off the condensed moisture from time to time. The mercury will be found to stand at the freezing point till all the ice is melted, the water having absorbed or rendered *latent* the heat supplied. It is this latent heat that is given out when water solidifies.

Water boils at the level of the sea at 100° C or 200°F; but if the pressure be reduced, ebullition takes place at a lower temperature.

**74.** The other compound of oxygen with hydrogen is **Hydroxyl,** a body of great theoretical interest to the chemist, as it occurs in all the oxy-acids. Its formula is $\left\{ \begin{matrix} OH \\ OH \end{matrix} \right\}$ but abbreviated the molecule is written $Ho_2$, and the semi-molecule or atom Ho. It is hence a monad radical, and is capable of being replaced by Nao, Ko, and its molecule by the dyad radicals Zno″, Cao″, etc. It was known formerly as oxygenated water, or peroxide of hydrogen. It can be prepared by passing a current of carbonic acid gas through water containing baric peroxide $BaO_2$. Baric carbonate is produced, and the atom of oxygen liberated combines with the water.

$$\left\{ \begin{matrix} O \\ O \end{matrix} \right. Ba'' + CO_2 + OH_2 = COBao'' + \left\{ \begin{matrix} OH \\ OH \end{matrix} \right. \text{ or } H_2O_2 \text{ or } HO_2$$

It may also be prepared by adding HCl to the water in which the baric peroxide is suspended, afterwards adding cautiously sulphuric acid to throw down the barium as sulphate.

$$BaO_2 + OH_2 + 2HCl = BaCl_2 + OH_2 + H_2O_2$$
$$\text{Then } BaCl_2 + SO_2Ho_2 = SO_2Bao'' + 2HCl$$

Hydroxyl is a colourless liquid of syrupy consistence, and prepared in either of the above ways appears as a heavy liquid distinct from the water. It has an odour something like that of chlorine. Its sp. gr. is 1·5. It must be concentrated by evaporating over sulphuric acid under the receiver of the air-pump. A solution of it may be purchased at 372 Oxford Street, London, and no doubt at other places, being used as medicine under the name of peroxide of hydrogen.

*Exp.* 9. Heat some of the solution of hydroxyl in a test-tube. Oxygen is evolved, which will rekindle a glowing splinter.

*Exp.* 10. Blacken a lead paper with sulphuretted hydrogen and put in the solution; it will be bleached. Clean also some discoloured white paint with the solution. It is an unstable compound, and a powerful oxidising agent. It is used on this account for cleaning oil paintings.

$$PbS + 4Ho_2 = SO_2Pbo'' + 4HO_2$$

Metallic silver is oxidised by it, but if argentic oxide be suspended in water, and hydroxyl added, both are decomposed.

$$Ag_2O + H_2O_2 = Ag_2 + OH_2 + O_2$$

It liberates iodine from potassic iodide.

$$2KI + H_2O_2 = 2OKH + HI_2$$

Hydroxyl is transformed to water by the action of nascent hydrogen (that is, just as it is generated).

*Exp.* 11. Generate a little hydrogen in a test-tube by the action of dilute HCl on zinc. Add some of the solution of hydroxyl, the evolution of gas ceases. The same takes place when $H_2O_2$ is added to the voltameter in the electric decomposition of water. Hydrogen ceases to be given off.

$$SO_2Ho_2 + Zn + H_2O_2 = SO_2Zno'' + 2OH_2$$

---

## CHAPTER XIII.

### Compounds of Chlorine with Oxygen and Hydroxyl.

75. Chlorine and oxygen do not combine directly with each other, and compounds of these elements are therefore

only obtained by indirect means. They are, moreover, very *unstable*, many of them decomposing at low temperatures with explosion. The author would refer to Dr Frankland's Lecture Notes for a complete list of these compounds, as many of them are only of theoretical importance, and a mere description of them would be beyond the scope of this elementary treatise. The oxides or anhydrides are converted into acids by the substitution of an atom of H, or the monad radical hydroxyl, (Ho) for an atom of chlorine. Thus hypochlorous anhydride Cl – O – Cl or $OCl_2$ by substitution of an atom of H becomes hypochlorous acid, Cl — O — H or ClHo (OClH). This is effected by the action of water upon these bodies, as will be seen below.

*Exp.* 1. Pass dry chlorine gas (dried by chloride of calcium) through a tube about six inches long containing mercuric oxide, which must be kept cool. Connect this with a test tube surrounded by a freezing mixture (ice and salt), a yellowish gas will pass over which will condense to a deep red liquid, emitting a vapour of a deeper colour than chlorine. It is hypochlorous anhydride, mercuric oxychloride being also formed.

$$2HgO + 2Cl_2 = \left\{\begin{matrix} HgCl \\ O \\ HgCl \end{matrix}\right. + OCl_2$$

The liquid $OCl_2$ is highly explosive, decomposing *by the heat of the hand* into chlorine and oxygen.

*Exp.* 2. Disconnect the test-tube in the last experiment, but let it remain in the ice. Add some water to it; it will be converted into hypochlorous acid.

$$OCl_2 + OH_2 = 2ClHo$$

The same reaction takes place when chlorine is passed into water containing mercuric oxide in suspension, or when mercury is shaken up in a test-tube with the gas.

$$2HgO + OH_2 + 2Cl_2 = \left\{\begin{matrix} HgCl \\ O \\ HgCl \end{matrix}\right. + 2ClHo$$

Hypochlorous acid is a powerful oxidising agent, giving off chlorine when exposed to the light.

*Exp.* 3. Add a few drops of HCl to the ClHo in last experiment;

chlorine will be evolved from both compounds. It will bleach, and not merely redden, a litmus paper.

$$ClHo + HCl = OH_2 + Cl_2$$

If argentic oxide be added, oxygen is evolved from both compounds.

$$OAg_2 + 2OHCl = 2AgCl + OH_2 + O_2$$

Metallic oxides and hydrates are converted into salts called hypochlorites, acids terminating in *ous* forming salts ending in *ite*. The monad radical sodoxyl replacing hydroxyl.

$$ONaH + ClHo = ClNao + OH_2$$

It was supposed at one time that *chlorides* and *hypochlorites* were both formed when chlorine gas was made to act upon certain metallic oxides and hydrates, as in the manufacture of *bleaching powder*, or the so-called "chloride of lime," *i.e.*,

$$2CaHo_2 + 2Cl_2 = CaCl_2 + \underset{\text{Calcic hypochlorite.}}{CaClO_2} + 2OH_2$$

But this so-called "chloride of lime" does not contain calcic chloride when properly made. Calcic chloride being deliquescent and soluble in alcohol, which is not found to be the case with good bleaching powder.

The following appears to be the true reaction :—

$$Cl_2 + Ca\ Ho_2 = \underset{\text{Calcic chloro-hypochlorite.}}{Ca(OCl)Cl} + OH_2$$

The graphic formula of bleaching powder, or calcic chloro-hypochlorite, is Cl – Ca – O – Cl. The hypochlorites readily give up oxygen to decaying animal and vegetable substances and also to colouring matters containing hydrogen, and hence their use as disinfecting and bleaching agents.

Bleaching powder readily yields chlorine when acted upon by weak acids, even the carbonic acid gas in the air being sufficient to evolve it slowly from "chloride of lime."

$$Ca(OCl)Cl + SO_2Ho_2 = SO_2Cao'' + OH_2 + Cl_2$$
$$Ca(OCl)Cl + 2HCl = CaCl_2 + OH_2 + Cl_2$$
$$Ca(OCl)Cl + CO_2 = COCao'' + Cl_2$$

**76. Chloric Peroxide** Cl–O–O–O–O–Cl $\left\{\begin{array}{l}OCl\\O\\O\\OCl\end{array}\right.$ or $Cl_2O_4$.

This is a gas at ordinary temperatures, but by slight pressure, or by exposure to a cold of 20°C, it can be reduced to a red liquid, which explodes violently. The gas is of a deep colour, and has an irritating odour. It decomposes slightly above the ordinary temperature of the air, or on exposure to sunlight, and in contact with certain bodies. It is prepared from potassic chlorate.

*Exp.* 4. Put a few crystals of chlorate of potash into a test-tube, add a few drops of sulphuric acid—chloric peroxide, $Cl_2O_4$ will be evolved. Make a copper wire hot, and plunge it in the gas, it explodes without danger.

Sulphuric acid should be heated very cautiously with potassic chlorate, as the decomposition of the $Cl_2O_4$ causes a series of explosions, making a continuous crackling noise, frequently projecting the materials out of the tube. The only safe way to heat it is in a hot-water bath.

*Exp.* 5. Fill an ale-glass or tall jar with water, and put it upon a plate. Drop a small chip of phosphorus into the glass, and cover it with crystals of potassic chlorate. Pour some sulphuric acid on to the chlorate by means of a long thistle-headed funnel. The potassic chlorate will be decomposed, and chloric peroxide given off, which decomposes immediately in contact with the phosphorus with a crackling noise, flashes of green light appearing under water, and a greenish-coloured gas, having a sweet odour, being evolved and dissolving in the water. The decomposition of the potassic chlorate gives potassic perchlorate, hydric potassic sulphate, water, and chloric peroxide :

$$3\left\{\begin{array}{l}OCl\\OKo\end{array}\right. + 2SO_2Ho_2 = \left\{\begin{array}{l}OCl\\O\\OKo\end{array}\right. + 2SO_2HoKo + OH_2 + \left\{\begin{array}{l}OCl\\O\\O\\OCl\end{array}\right.$$

*See* also *Exp.* 5, Chap. I. (*violet flame*). Baric chlorate might be substituted with the production of a *green flame*.

*Exp.* 6. Rub a crystal of potassic chlorate in a mortar with a bit of sulphur; combustion of the sulphur takes place, with explosion. Take care to operate only on a small quantity.

Lucifer matches are tipped with a mixture of potassic chlorate, phosphorus, and gum, and the crackling noise made when they are rubbed on the sand-paper is due to the cause stated above.

**77. Preparation of Potassic Chlorate** $\left\{ \begin{matrix} OCl \\ OKo \end{matrix} \right.$

This may be prepared by passing chlorine through a concentrated hot solution of potassic hydrate. Potassic chlorate and potassic chloride are both formed, but the latter being much more soluble than the former, the *chlorate* crystallises out and the other is left in solution. The chlorate should be dissolved and re-crystallised to purify it.

$$3Cl_2 + 6KHo = \left\{ \begin{matrix} OCl \\ OKo \end{matrix} \right. + 5KCl + 3OH_2$$

On account, however, of the expense of potassic hydrate, milk of lime is used instead, whereby calcic chlorate and chloride are produced, potassic chloride being afterwards added to the mixture; double decomposition takes place, calcic chloride and potassic chlorate being formed.

*Exp.* 7. Pass chlorine to saturation (till the liquid evolves the gas) into milk of lime ($CaHo_2$) kept boiling in a beaker.

$$6CaHo_2 + 6Cl_2 = \left\{ \begin{matrix} OCl \\ O \\ Cao'' \\ O \\ OCl \end{matrix} \right. + 5CaCl_2 + 6OH_2$$

Filter and add a solution of potassic chloride to the mixture; potassic chlorate and calcic chloride will be formed. The calcic chloride being very soluble in water, the potassic chlorate will separate on cooling. The "mother liquid" being drained off, the chlorate should be re-crystallised, as above

$$\left\{ \begin{matrix} OCl \\ O \\ Cao'' \\ O \\ OCl \end{matrix} \right. + 2KCl = CaCl_2 + 2 \left\{ \begin{matrix} OCl \\ OKo \end{matrix} \right.$$

The acid, of which potassic chlorate is a salt, is **Chloric acid,** its formula being H—O—O—O—Cl

$$\left\{ \begin{matrix} OCl \\ OHo \end{matrix} \right. \text{ or } \left\{ \begin{matrix} OCl \\ O \\ OH \end{matrix} \right.$$

The acid itself, however, is unimportant. It may be prepared from baric chlorate by the action of sulphuric acid. The *chlorates* are all soluble, and cannot therefore be precipitated from their solutions. They give up their oxygen more readily than the nitrates, which are somewhat similar in their properties.

If potassic chlorate be heated till one-third of its oxygen is given up, and the fused mass be allowed to cool, when dissolved in water a salt will crystallise out from the chloride, which contains one atom of oxygen more than the chlorate. It is potassic perchlorate, and has the formula $KClO_4$

$$2\left\{\begin{matrix}OCl\\OKo\end{matrix}\right.=KCl+\left\{\begin{matrix}OCl\\O\\OKo\end{matrix}\right.+O_2$$

This occurred as a secondary product in making chloric peroxide, and was mentioned also under the head of oxygen. From this, **perchloric acid** $\left\{\begin{matrix}OCl\\O\\OHo\end{matrix}\right\}$ is prepared by the action of sulphuric acid. It is a powerful oxidising agent.

## CHAPTER XIV.

### BORON B‴.

**78.** THIS is a triad element. It is never found in a free state in nature. Its principal sources are—*tincal*, a crude kind of borax obtained from certain lakes in Thibet; and boracic, or boric acid, obtained from the lagoons in the volcanic district of Tuscany. It has also been found lately, in combination with soda, in the lakes of California and Utah—this being also crude borax or sodic borate ($B_4O_5Nao_2$). From borax, the tribasic boric acid ($BHo_3$) is prepared by the action of hydrochloric or sulphuric acids.

*Exp.* 1. Prepare a hot saturated solution of *borax* in a beaker. Add strong HCl till the solution reddens blue litmus paper. Crystals of boric, or boracic acid separate on cooling, sodic chloride being left in solution.

$$B_4O_5Nao_2 + 2HCl + 5OH_2 = 4BHo_3 + 2NaCl.$$

The crystals should be washed with water and spread

on a filter paper. They consist of flat plates having a pearly lustre and greasy to the touch. They are soluble in hot water and alcohol. Their solution in water imparts a port-wine tint to *litmus*. $BHo_3$ is a weak acid in the cold, and turns *turmeric* paper reddish brown, even when free HCl is present.

*Exp.* 2. Dissolve some of the crystals in methylated spirit. Inflame the spirit—the $BHo_3$ volatilises and imparts an apple-green colour to the flame. Borax may be dissolved, but requires the addition of HCl or $SO_2Ho_2$ to liberate the acid.

Boric acid will expel $CO_2$ from the solution of a carbonate on boiling.

*Exp.* 3. Make a saturated solution of carbonate of soda in a test-tube, and add some crystals of $BHo_3$, obtained in last experiment. Fit a cork and tube, and a bit of flexible tube, to lead into another test-tube containing *lime-water*. Heat to boiling; carbonic acid gas will be evolved, and turn the lime-water milky.

$$3CONao_2 + 2BHo_3 = 2BNao_3 + 3CO_2 + 3OH_2$$

By heating crystals of the tribasic boric acid to the boiling point, they lose water, and are converted into an acid having the composition BOHo, called *metaboric* acid.

$$BHo_3 = BOHo + OH_2$$

*Exp.* 4. Heat some of the crystals to redness on charcoal before the blow-pipe flame, they become anhydrous, and partly volatilise if the heat be continued. They are converted into boric anhydride $O{=}B{-}O{-}B{=}O$ or $B_2O_3$. If this boric anhydride be heated with sodium on charcoal before the blow-pipe, sodic oxide will be formed, and gritty, amorphous, brownish-black pieces of Boron will appear mixed in the mass.

$$B_2O_3 + 3Na_2 = 3ONa_2 + B_2$$

The element exists in two allotropic states:—

*a.* Amorphous. *b.* Graphitoidal or diamond boron. It combines directly with nitrogen, which few elements do, forming boric nitride $B'''N'''$. Also with chlorine, &c.

*Exp.* 5. Fuse in a porcelain crucible a mixture of two parts by weight of *sal-ammoniac* (ammonic chloride $NH_4Cl$) and one ounce powdered *borax*. Abundance of white fumes will be given off from HCl. A white mass will be left, consisting of NaCl and BN. Dissolve out the sodic chloride, and a white amorphous infusible powder will be left. For this, and other reactions, see Dr Frankland's Lecture Notes.

The most important compound of boron is borax. It is a borate of soda. Upon ignition, the crystals lose water and become anhydrous (refined borax). It is largely used for glazing pottery, forming a glass-like compound with metallic oxides. It is also used for soldering, on account of its property of dissolving the film of oxide, thus keeping the surfaces of the metals clean and enabling them to alloy with one another. The same property makes it of great use in *Analytical Chemistry,* the borax dissolving certain metallic oxides, which impart a particular colour to the glassy mass.

*Exp.* 6. Take a bit of platinum wire about 5 inches long, and turn it up at the end, forming a loop. Heat it to redness in the Bunsen flame, and then dip the loop into powdered borax. Heat the mass taken up before the blow-pipe, it swells up and fuses to a transparent bead. Whilst hot, dip the bead into a solution of nitrate of cobalt, and heat again strongly, the bead will be coloured *blue.* Salts of copper, iron, manganese, nickel, and chronium, impart their own characteristic colours to borax beads. The chlorides are best for the purpose, or nitrates.

---

# CHAPTER XV.

## CARBON $C^{IV}$ AND ITS COMPOUNDS WITH OXYGEN.

79. CARBON is one of the *tetrad* elements. One of its compounds with hydrogen has been already alluded to, but such compounds, which are very numerous, form a separate branch of the subject under the head of Organic Chemistry, sometimes called the chemistry of the carbon compounds.

Carbon exists in three allotropic states: *crystallised,* and almost pure in the *diamond,* with a sp. gr. of 3·3; in *flat tabular plates,* as in *graphite* or black-lead, sp. gr. 2.15; the black scales found in pig-iron consist of this variety of carbon; and, *amorphous,* as in the different kinds of *charcoal.* Coal, coke, and lamp-black also contain hydrogen, oxygen, and earthy matters.

Wood charcoal is made by burning billets of wood made into heaps, covered with clay to prevent access of air except in a few places. The inflammable gases burn away, and charcoal is left. Or, it is heated in iron cylinders, the gases and vapours evolved being made to pass through a pipe into receivers, where they are condensed.

*Exp.* 1. Light a splinter of wood, and plunge it into a test-tube. It will only burn with a flame at the mouth of the tube. The part inside will glow, and at last become a charred mass. The wood consists principally of C, H, and O; the two latter burn away and leave the carbon. Brown tarry matters condense on the sides of the tube. If small pieces of wood are heated in a test-tube, to which a glass delivery-tube is adapted, the gases and vapours may be condensed—among which may be recognised tar and pyroligneous acid, which will redden litmus.

Wood may be heated in a Cornish crucible, loosely covered with a bit of earthenware, with the same result. Carbon is infusible.

Animal and vegetable charcoal are very porous, and have the power of absorbing gases.

*Exp.* 2. Collect a jar of dry hydrochloric acid gas over mercury, and plunge a piece of heated charcoal under the jar. The gas will be absorbed, and the mercury rise in the jar. See also *Exps.* 4 and 5, page 81. Mercury only being used.

Charcoal absorbs 90 times its volume of ammonia gas, 85 HCl, 55 of sulphuretted hydrogen, nearly 10 of oxygen, and 2 of hydrogen.

It is on this account used as a disinfectant; the best explanation of the process being, that oxygen is condensed in the pores of the charcoal, and the putrescent substances become oxidised.

*Exp.* 3. Shake up a piece of tainted meat in a test-tube containing water and some finely powdered charcoal, the smell will disappear. Stagnant water may be deodorised in the same way.

**80.** Animal charcoal has the power of absorbing colouring matters, and many bitter principles when in a finely divided state, in addition to its power of absorbing gases. On this account it is used for *filters*, and in the decolourisation of sugar in sugar refining.

*Exp.* 4. Get some dark brown sugar (foots) and make a strong solution of it. Put some animal charcoal or bone-black into a large boiling-tube, and pour the solution of sugar into it. Shake it well and pour it upon a filter, a pale, straw-coloured liquid will run through. If the process be repeated, the whole of the colour will be retained by the charcoal, but the syrup loses none of its sweetness.

Port wine acted on in the same manner loses its colour and also its *aroma*.

**81. Oxygen Compounds of Carbon.** When any kind of carbon is burnt in oxygen, carbonic anhydride, usually termed carbonic acid gas, is produced. Thus, when a small splinter of *diamond* is suspended on a bit of platinum foil in a jar of oxygen, and an electric current, sufficiently powerful to heat the foil—and hence the diamond—to redness, it glows and swells up like a mass of coke, *carbonic anhydride* $CO_2$ is produced, and a very small quantity of *ash*, or earthy matter, is left.

A piece of graphite from a lead pencil heated to redness in a Bunsen flame, and plunged into oxygen, also produces $CO_2$.

Lime-water poured into the jars in either case is turned milky from the production of chalk or calcic carbonate.

$$CaHo_2 + CO_2 = COCao'' + OH_2$$

*Exp.* 5, Chap. XI., may be repeated here. No change of volume takes place when carbon is burnt in oxygen. Carbonic anhydride is prepared by the action of HCl, $SO_2Ho_2$, or vinegar upon a carbonate. HCl is the best to use, as the chloride of calcium formed is more soluble than the sulphate, which is apt to form a coating over the substance, and prevent the further action of the acid.

*Exp.* 1. Put into four beakers some old mortar, powdered oyster shells, limestone or chalk, and pounded coral. Pour upon each of them some HCl or dilute sulphuric acid. An invisible gas will come off, which will extinguish a lighted taper lowered into it. The carbonic anhydride evolved is a non-supporter of combustion.

*Exp.* 2. Break a piece of white marble—calcic carbonate in a crystalline state—into small pieces, and put it into the hydrogen-generating bottle. Cover it with water, and pour on it some HCl. A violent effervescence will take place, and $CO_2$ will be evolved. It may be collected over water in the pneumatic trough as usual, although *water dissolves its own volume of the gas.*

The water is put into the generating bottle to dissolve the calcic chloride formed. Shells, chalk, and coral consist principally of calcic carbonate, and the action of acids upon them is the same as upon marble—but marble is preferable, as it does not froth up so much.

$$COCao'' + 2HCl = CaCl_2 + OH_2 + CO_2$$

Collect several jars or bottles of the gas in this way. It may also be collected by *downwards displacement*, as in the case of chlorine.

*Exp.* 3. Plunge a lighted taper into a jar of the gas, it will be extinguished.

*Exp.* 4. Fill a bottle with water, and then half fill it with $CO_2$. Cover it with the hand and shake it well. The hand will be pressed in, showing that the gas has been dissolved. Open it in trough, water rushes in. Pour in some lime-water, it turns milky, but upon agitation it re-dissolves. Chalk or calcic carbonate, though insoluble in water, is soluble in water containing carbonic acid gas. This enables our sea and river waters to hold chalk in solution.

*Exp.* 5. Boil some of the water containing the dissolved chalk, it will be re-precipitated.

*Exp.* 6. Pour a solution of litmus into a jar of carbonic anhydride, it is turned wine-red. Carbonic acid has never been isolated, but the dry gas would not affect a dry litmus paper.

**82.** Carbonic anhydride is a very heavy gas, its density compared with hydrogen being 22. A litre of the gas would weigh ·0896 × 22 or 1·9712 grams. Its weight compared with air is therefore $\frac{22}{14·47}$ or 1·529.

*Exp.* 7. Fill a large bell-jar with $CO_2$ by downwards displacement, the heavy gas lifting out the air. Cover with a glass plate all but the opening left for the delivery-tube. The gas may be tested from time to time by a lighted taper. When full, the taper will be extinguished at the opening, as the jar will run over.

Tie a string round a small beaker to act as a pail, and bale out a beaker of the gas, which pour on to a lighted taper.

*Exp.* 8. Make some lather, and blow soap-bubbles, letting them fall into the large bell-jar of $CO_2$. They will float on account of the greater lightness of the contained air.

The atmosphere contains about 4 parts in 10,000 or ·04 per cent. of carbonic acid gas. This is largely increased where great numbers of persons are congregated together, unless there be a good current of air. Even 2 per cent in the air produces discomfort, 5 per cent. or less faintness, and 10 per cent. would be absolutely fatal. Air that is exhaled from the lungs contains 5 per cent. more $CO_2$ than that which is inhaled. This is produced from the chemical union of *waste matters* with the oxygen of the air. This is constantly being thrown off from the blood.

*Exp.* 9. Leave a saucer containing *lime-water* exposed to the air, a crust of carbonate of lime or chalk will be formed upon it.

*Exp.* 10. Breathe through a glass tube into lime-water. It turns milky from the $CO_2$ contained in the breath.

A few drops of HCl dissolves the calcic carbonate formed.

Air drawn through lime-water into the lungs does not affect it, owing to the smallness of the quantity of $CO_2$. If the air drawn in be sent into another vessel also containing lime-water, as in *Exp.* 10, it turns milky. An ingenious bit of apparatus for showing this experiment is shown in Fig. 37. It consists of two small flasks, each provided with corks fitted with two tubes. One tube in *a* passes into the lime-water, the other just passes through the cork, and is connected with a flexible tube leading to the mouth, provided with a valve of oiled silk opening inwards. The mouth-piece is also connected with the tube, having a valve opening outwards, which enters the lime-water in *b*, and when air is driven out of the lungs it turns it milky. Several full inspirations and expirations being taken, the effect is very marked.

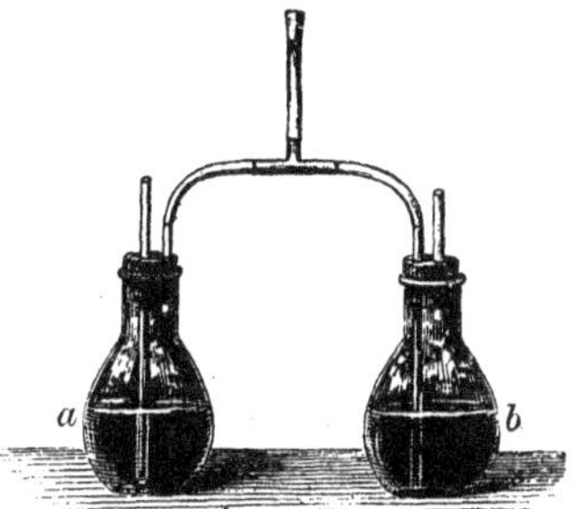

Fig. 37.

*Exp.* 11. Burn a taper in a bottle till it goes out. Then pour in lime-water, it will turn milky. $CO_2$ is produced in the burning of our lamps, gas, candles, and fires.

It is also produced in all cases of fermentation. Effervescent drinks are produced by the action of some acid on a carbonate. And in consequence of the solubility of the gas, ærated waters are made by forcing the gas into them by great pressure. The gas, on account of its weight, collects in brewers' vats. It is also found in old wells and caverns. A lighted candle should be lowered into such places before a man is allowed to descend, as fatal accidents may occur if this precaution be neglected.

Limestone or chalk, when heated to redness in a kiln, give off $CO_2$, and become calcic oxide or quick-lime.

Magnesium wire ignited and plunged into a jar of $CO_2$ decomposes it, and carbon is deposited. If dry $CO_2$ be sent through a bulb-tube containing potassium heated to redness it is likewise decomposed.

If, however, carbonic anhydride be passed through an iron tube containing charcoal, heated to redness in a furnace, it will be decomposed and take up another atom of carbon, forming a gas with one half the oxygen called carbonic oxide CO. Or,

*Exp.* 12. Pass a stream of $CO_2$ over powdered charcoal contained in a hard glass tube about ten inches long, heated to redness by Herepath's blow-pipe. The carbonic oxide will burn at the end of the tube with a beautiful blue flame.

$$CO_2 + C = 2\,CO$$

This gas is often seen burning at the top of a clear fire. Carbonic acid is produced by the burning coals at the bottom of the grate, and in passing through the red hot fire is decomposed, and takes up another atom of carbon. The CO burns at the top of a *clear fire* with a blue flame, forming $CO_2$. If our fires could be fed from the bottom, and brilliant combustion produced, so that the carbon could be heated to redness, very little smoke or unburnt carbon would be produced. It is on this principle furnaces are constructed to consume their own smoke, great saving of the fuel being effected. Be careful not to inhale carbonic oxide, as it acts as a direct poison.

*Exp.* 13. Heat some bits of chalk with charcoal in a test-tube of hard glass by means of the Herepath. The carbonate is decomposed and CO produced.

**83. Carbonic oxide** is usually prepared by decomposing oxalic acid by means of sulphuric acid, which abstracts the elements of water from this compound, $CO_2$ and CO are both formed.

*Exp.* 14. Put some oxalic acid into a flask, and pour on it some $SO_2Ho_2$. Heat gently, and convey the mixed gases through lime-water or solution of caustic soda in a Woulff's bottle. The $CO_2$ will form calcic or sodic carbonate, and the carbonic oxide may be collected over water as usual. Collect two jars of the gas.

*Exp.* 15. Inflame a jar of CO, and immerse a lighted taper in the gas, it will be extinguished.

*Exp.* 16. Pour some lime-water into a jar of CO and agitate it, no milkiness is produced. Inflame the gas and then shake it again, the lime-water will turn milky. The CO combines with the oxygen of the air and forms $CO_2$ again. This takes place at the top of the clear fire as mentioned above. When burnt it forms an equal volume of $CO_2$.

Carbonic oxide may be prepared by adding sulphuric acid to ferro-cyanide of potassium (yellow prussiate of potash). Care must be taken, however, as ferro-cyanic or prussic acid may be produced at the same time.

The sp. gr. of CO compared with H is $\frac{12+16}{2}=\frac{28}{2}$ or 14; and with air ·9724. A litre of the gas weighs 1·254 grams.

*Question* 1. How much $CO_2$ can be obtained from 100 grams of calcic carbonate by the action of HCl?

*Ans.* 44 grams or 22·320 litres.

2. What is the weight of 10 litres of carbonic acid gas compared with hydrogen as unity?

*Ans.* 19·7120 grammes.

---

## CHAPTER XVI.

### NITROGEN—THE ATMOSPHERE—COMBUSTION FLAME—THE BLOW-PIPE.

**84.** NITROGEN $N^v$ is a pentad element. It forms $\frac{4}{5}$ of the atmosphere, which is a mechanical mixture of this gas with oxygen and a little carbonic acid and watery vapour in varying proportions. It may be obtained from the air in various ways by withdrawing the oxygen.

*Exp.* 1. Graduate a stoppered bell jar, dividing it into five equal portions by pasting a slip of paper down the side. Put a bit of phosphorus in a small evaporating dish and float it upon water. Ignite the phosphorus and cover it with the jar. The phosphorus will burn at the expense of the oxygen of the air contained in the jar, forming $P_2O_5$* When the contents of the jar cool the water will rise and occupy one-fifth of the space. The remaining four-fifths of the jar contain **nitrogen.**

Fig. 38.

On plunging a lighted taper into the gas it will be extinguished. Nitrogen is a non-supporter of combustion. It has no action upon test-papers, it is therefore a neutral body. It does not turn lime-water milky.

*Exp.* 2. Repeat the last experiment, and then restore the fifth part with which the phosphorus combined by sending oxygen into the jar. A candle will burn in the mixture as it does in air, so that air is not a chemical compound.

*Exp.* 3. Shake up a measured quantity of air in a test-tube with a solution of *pyrogallic acid* in strong caustic potash. This absorbs the oxygen. That which remains will be *nitrogen.*

The volumetric composition of air may be proved by exploding a measured volume with hydrogen in the eudiometer mentioned in *Exp.* 1, Chap. XII., but such experiments are difficult for students. Nitrogen may also be obtained by passing air over red-hot copper; the metal unites with the oxygen, leaving the nitrogen. Nitrogen may be obtained from several of its compounds.

*Exp.* 4. Heat in a 2-oz. flask some of the salt called ammonic nitrite (formed by the action of *nitrous* acid upon ammonia—shown here as the oxide of the body ammonium $NH_4$). It splits up into nitrogen, which is evolved, and water. The gas can be collected over water as usual.

$$NO\,(NH_4O) = 2OH_2 + N_2$$

* Keep the hand on the jar till the P. goes out.

*Exp.* 5. Heat in a small flask a mixture of ammonic chloride and *sodic* or potassic *nitrite.* Nitrogen is given off.

$$NH_4Cl + NONao = NaCl + 2OH_2 + N_2$$

**85.** Nitrogen can also be obtained by passing chlorine gas through an excess of strong solution of ammonia contained in a three-necked Woulff's bottle. The ammonia *must be in excess,* or a terribly explosive compound called chloride of nitrogen $NCl_3$ will be formed in brown drops by the combination of some of the chlorine with the nitrogen of the ammonia. The safety-tube, which must be a funnel in this case, must be constantly smelt to ascertain whether free ammonia be present. The reactions are given below.

*It is not advisable for a student to try the experiment.*

$$8NH_4Ho + 3Cl_2 = 6NH_4Cl + N_2 + 8OH_2$$

Dangerous. $NH_4Ho + 3Cl_2 = NCl_3 + OH_2 + 3HCl$

**86.** Nitrogen does not combine readily with other elements at ordinary temperatures. When a flash of lightning passes through air, it causes some of the oxygen and nitrogen to combine, and with moisture to form nitric acid. When electric sparks are passed through air, the same result is produced, as may be shown by the action on blue litmus paper, which will be reddened.

The composition of the air is very constant, varying but little, whether taken from the mountain top, or at the sea level. The air of towns contains various impurities, and all air contains watery vapour, together with a small quantity of carbonic acid gas which is being constantly exhaled by animals, and absorbed by the leaves of plants.

*Percentage composition of the atmosphere.*

| | By volume. | By weight. |
|---|---|---|
| Nitrogen, . . . | 79·1 | 76·9 |
| Oxygen, . . . . | 20·9 | 23·1 |
| | 100· | 100· |

We cannot do better than introduce at this stage a few remarks on the phenomena of combustion.

**87. Combustion** takes place when *chemical action* is sufficiently violent to heat bodies to redness. We usually call the body so heated *a combustible*, and the gas or vapour in which it burns, the *supporter of combustion.*

Hence when hydrogen is burnt in air or oxygen, it is usual to say hydrogen is the combustible body and oxygen the supporter of combustion. In the same manner we say sulphur is combustible when burnt in oxygen. But, if sulphur be heated in a flask till it gives off vapour, copper leaf will burn in it, the sulphur being in this case the supporter of combustion. A jet of hydrogen will burn in chlorine, the latter being the supporter of combustion, according to the ordinary phraseology, but it would not be difficult to show that chlorine will burn in hydrogen, when chlorine would appear to be the combustible body and hydrogen the supporter of combustion. **Combustion,** however, is in fact nothing more than **chemical combination at a great heat.**

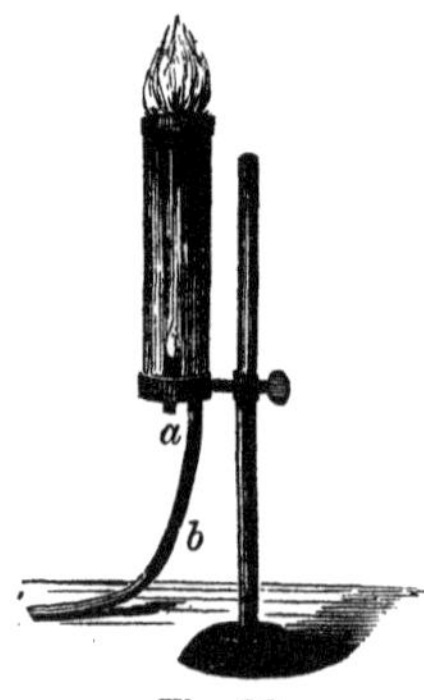

Fig. 39.

*Exp.* 6. Get an ordinary gas chimney; fit a large cork to the end, perforated with two holes for glass tubes, as in Fig. 39, which is a similar arrangement, only having a metal base. Fix it by a clip to a stand, and put a bit of fine wire gauze at the top of the cylinder. Close the tube *a* with a cork, let gas in to the bottom of the cylinder by the tube *b*, and light it at the top of the wire gauze. Coal gas will of course fill the cylinder. Now pass a lighted match through *a*, and a shell of flame appears at the top of the tube, the air appears to burn in the coal gas. The breath from the mouth may be substituted for the air by attaching a piece of flexible tube to *a*. The air and gas burn really only where they come in contact with one another.

For bodies to burn, they must first be raised to a certain temperature. This is called the igniting point. In some cases a moderate heat is sufficient to set up combustion; the friction on the sand-paper being sufficient to inflame the phosphorus on the lucifer match. Some bodies require intense heat to make them burn, whilst zinc ethyl burns at the ordinary temperature of the air, and carbonic disulphide when a hot glass rod is brought near it. In the latter case, as in many others, the substance gives off vapour, and it is this which inflames.

*Exp.* 7. Pour a little ether into an evaporating dish and bring a light near it. The ether gives off vapour at the ordinary temperature of the air, and this inflames.

Alcohol and paraffin must be warmed gently before sufficient vapour is produced to inflame. Some kinds of the latter, however, give off vapour at a very low temperature, and are hence dangerous.

The ordinary candle flame is produced by the burning wick converting the wax or tallow into vapour, and it is this which inflames.

**88. Flame** is produced by burning gas or vapour, and the more intense the agitation the brighter is the flame. Hence the flame of coal gas, which consists of a mixture of gases and vapours, is greatly intensified by a jet of oxygen being introduced into it, the combustion being more vivid.

The conical form of a flame is caused by ascending currents of air. The flame of a candle consists of three concentric cones. The *outer one,* which is scarcely perceptible, is *the area of complete combustion,* where the oxygen of the air comes in contact with the burning gases and vapours; it is sometimes called the *oxidising flame.* The next or middle shell is the *area of partial combustion* sometimes called the *deoxidising flame.* The *centre* of the flame is *hollow,* consisting of the unburned gases.

*Exp.* 8. Put a piece of wood across a candle flame, it will be charred on each side of a central part, which is not discoloured; or, lower a card upon the flame, a black ring will be produced.

*Exp.* 9. Put some gunpowder on the wire gauze at the top of the cylinder in *Exp.* 6. Then light the gas. The gunpowder will not inflame till the gas is lowered considerably, showing that the flame is hollow; or allow gas to pass through a bit of wire gauze, while a light is applied above, the flame does not pass through. The gauze cools the flame. This is the principle of Davy's safety lamp.

The Bunsen flame is not hollow, the air mixing with gas as it passes up the tube.

**The blow-pipe flame** consists of two parts only, as air is blown into the middle of the flame. The outer is called the *oxidising*, and the inner the *deoxidising* flame, as above.

Familiarity with the use of the blow-pipe may be obtained by trying the following experiments :—

*Exp.* 10. Put some red lead in a little hollow, scooped in a bit of charcoal. Heat strongly just at the point of the inner cone. The $Pb_3O_4$ will be reduced, and a bead of metallic lead will appear. Heat this bead in the outer flame and a yellowish-red *incrustation* will appear on the charcoal, the metal being oxidised again.

*Exp.* 11. Heat some sulphate of zinc on charcoal, and a yellow incrustation will appear, which becomes white when cold. It consists of ZnO.

*Exp.* 12. Mix some cupric chloride with carbonate of soda, and heat strongly in the inner flame; red scales, consisting of the metal copper, will be obtained. The $CONao_2$ acts as a flux, and renders the copper salt more easily fusible.

*Exp.* 13. Mix some oxide of tin with potassic cyanide, or a mixture of carbonate of soda and potassic cyanide (fusion mixture), and heat on charcoal in the inner blow-pipe flame. The SnO will be reduced, globules of metallic tin will be produced with a slight incrustation if brought into oxidising flame.

---

## CHAPTER XVII.

### Oxides and Oxy-Acids of Nitrogen.

**89.** Nitrogen forms five distinct compounds with oxygen, and two of these are converted into acids by the addition of water; or, the latter may be considered as compounds of certain oxides with hydroxyl.

The most important compound of nitrogen is nitric acid. This is prepared from either of two of its compounds called potassic nitrate (nitre or saltpetre), and sodic nitrate, called also cubic nitre or Chili saltpetre.

Potassic nitrate, $NO_2Ko$—one of the ingredients in gunpowder—is found as an efflorescence on the soil in India, and nitrate of soda $NO_2Nao$, which cannot be used for manufacture of gunpowder as it is deliquescent, in a similar state in Chili and Peru. Either of these salts, heated with sulphuric acid, are decomposed, sulphates of the metals being formed, and nitric acid distils over.

*Exp.* 1. Heat some potassic nitrate in a small stoppered retort (Clark's retort, mentioned in Chap. IV., will do well for the purpose) with some strong sulphuric acid, keeping the receiver cool, as in Fig. 40. Orange-coloured vapours will be produced at the beginning, and again towards the close of the experiment, arising from the decomposition of some of the nitric acid produced. A pungent-smelling, fuming liquid *will distil over*, with a yellow tinge from the solution of some of the above-mentioned vapours.

Fig. 40.

A blue litmus paper will be instantly reddened.

One of the two atoms of hydroxyl in $SO_2Ho_2$ is exchanged for potassoxyl, hydric potassic sulphate being left as a white mass in the retort, whilst nitric acid $NO_2Ho$ will distil over.

$$NO_2Ko + SO_2Ho_2 = SO_2HoKo_2 + NO_2Ho$$

In the manufacture of nitric acid upon a large scale, the materials are put into iron cylinders lined with fire-clay which are heated in a furnace. On account of the great heat which can be obtained by this method, one-half the sulphuric acid need only be used, the neutral sulphate $SO_2Ko_2$ being produced. The heat, however, destroys some of the nitric acid. The acid is colourless when pure, but the commercial acid is yellow and contains iron and traces of sulphuric acid.

*Exp.* 2. Heat some colourless nitric acid in a test-tube, orange-red fumes will be produced by the decomposition of the acid—oxygen being liberated, and peroxide of nitrogen formed.

$$4NO_2Ho = 2OH_2 + O_2 = 2N_2O_4$$

**90.** Nitric acid is *monobasic*, and therefore forms but one class of salts, there being no *acid* **Nitrates.**

Nitrates are formed by the action of nitric acid upon metallic oxides and hydrates, and upon certain metals. They are all soluble, and are all decomposed by heat, the oxides of the metals being mostly formed.

Nitric acid is used as a solvent for many metals. It is a powerful oxidising agent. Thus, *tin* acted on by concentrated nitric acid withdraws $\frac{1}{5}$ of the oxygen, with the production of nitric peroxide $N_2O_4$; *silver*, in cold dilute acid, abstracts $\frac{2}{5}$ of the oxygen, with evolution of a red vapour of nitrous anhydride $N_2O_3$; *copper or mercury* withdraws $\frac{3}{5}$ of the oxygen, with the production of nitric oxide $N_2O_2$; and zinc, under proper conditions, removes $\frac{4}{5}$ of the oxygen, with the production of nitrous oxide $N_2O$. The body nitric anhydride $N_2O_5$, which added to water is transformed into nitric acid, is usually obtained by passing dry chlorine over argentic nitrate.

*Exp.* 3. Pour some of the nitric acid obtained in *Exp.* 1 upon some Dutch metal in a test-tube; violent chemical action will take place, the metal dissolving.

*Exp.* 4. Pour some strong nitric acid upon *gold leaf* in a test-tube; no effect is produced.

*Exp.* 5. Pour some strong HCl on gold leaf; no effect.

*Exp.* 6. Add the two together; the gold dissolves,—the mixture of nitric and hydrochloric acids is called *aqua-regia*. The nitric acid oxidises the H of HCl, forming water; whilst the nascent Cl combines with the gold, forming auric chloride $Au\,Cl_3$. The liberation of the chlorine is shown by the following equation—

$$2NO_2Ho + 6HCl = N_2O_4 + 2OH_2 + 3Cl_2$$

*Exp.* 7. Put some nitre in an old flask, heat it strongly; it will melt. While in a fused state pour in some powdered charcoal, and take the lamp away. A brilliant deflagration will take place, which will most likely break the flask. It is well, therefore, to put a basin of water under it. Nitre and charcoal form two of the ingredients in gunpowder, the former oxidises the latter, forming $CO_2$.

**91. Nitrous Oxide** $N_2O$, or laughing-gas, is prepared by heating ammonic nitrate in a small retort. The salt splits up into water and nitrous oxide. The gas should be collected over hot water. Ammonic nitrate is formed by adding nitric acid to a solution of commercial ammonic carbonate till the solution is neutral to test-papers, evaporating the solution to dryness. The gas should not be inhaled, except under medical advice.

$$NO_2(NH_4O) = 2OH_2 + ON_2$$

*Exp.* 8. Heat some ammonic nitrate in the small Clark's retort, Fig. 40, and collect two jars of the gas. The whole of the salt will volatilise.

*Exp.* 9. Insert a glowing splinter in the gas, it will be rekindled. A lighted taper will burn very brilliantly in the gas as in oxygen. The gas is decomposed.

*Exp.* 10. Just ignite a bit of sulphur and plunge it in the gas, it will be extinguished. Thoroughly inflame the sulphur and it will burn in the gas with a peculiar flame, the gas being decomposed.

**92. Nitric Oxide** $\left\{\begin{matrix} NO \\ NO \end{matrix}\right.$ or $N_2O_2$ $\begin{matrix} N = O \\ | \\ N = O \end{matrix}$

This is a colourless gas. It is prepared by the action of nitric acid upon copper, with the production of cupric nitrate, water, and $N_2O_2$ as a gas.

$$3Cu + 8NO_2Ho = 3\left\{\begin{matrix} NO_2 \\ NO_2 \end{matrix}\right. Cuo'' + 4OH_2 + N_2O_2$$

*Exp.* 11. Put into a wide-mouthed flask or Woulff's bottle, furnished with a thistle-funnel and delivery-tube, some copper clippings. Cover them with water, and pour strong nitric acid down the funnel. Red fumes will soon fill the flask, caused by the mixture of the gas evolved with the contained air. These soon disappear, and a colourless gas comes over, which is to be collected over water in jars. It is nitric oxide.

*Exp.* 12. Moisten a roll of blue litmus paper with sodic hydrate, and plunge it into a jar of the gas. Red fumes rapidly form down the jar, reddening the litmus paper above but not below, till the red fumes fill the jar.

The nitric oxide rapidly combines with the oxygen of the air to form nitric peroxide $N_2O_4$ and nitrous anhydride $N_2O_3$. The latter is converted into nitrous acid by the action of moisture.

$$N_2O_2 + O_2 = N_2O_4$$
$$\text{or } N_2O_2 + O_2 = 2N_2O_3$$
$$\text{and } N_2O_3 + OH_2 = 2NOHo$$

Nitric oxide is a test for oxygen, ruddy fumes being

produced as above when the two gases come together. This distinguishes oxygen from nitrous oxide, which rekindles a glowing splinter, but does not produce red fumes. Nitric oxide is completely dissolved by a solution of ferrous sulphate, forming a reddish liquid.

*Exp.* 13. Pass bubbles of oxygen into a jar of the gas standing over water in the pneumatic trough. The deep red gas produced will be speedily dissolved by the water which rises in the jar.

*Exp.* 14. Pass nitric oxide into a jar of oxygen, as in Exp. 11, Chap. XI., or into a jar of air.

*Exp.* 15. Plunge a burning taper into nitric oxide, it burns with a peculiar bright flame. A glowing splinter will not, however, be rekindled, the heat not being sufficient to decompose the gas.

*Exp.* 16. Just touch a bit of phosphorus with a hot wire in a deflagrating spoon, and plunge it quickly into the gas, it will be extinguished. If, however, the phosphorus be well alight before it is immersed in the gas, the heat will be sufficient to decompose it, and then the oxygen liberated will make it burn brilliantly.

**93. Nitrous anhydride** $N_2O_3$ is obtained by boiling nitric acid with starch or arsenious anhydride (white arsenic $As_2O_3$), a *deep-red* gas comes off, which condenses to a blue liquid in a bulb kept in a freezing mixture. It is also produced when silver is acted on by cold dilute nitric acid.

$$As_2O_3 + 2NO_2Ho = As_2O_5 + OH_2 + N_2O_3$$
$$6NO_2Ho + 2Ag_2 = 4NO_2Ago + 3OH_2 + N_2O_3$$

Liquid $N_2O_3$ with a small quantity of water produces nitrous acid NoHo, but with a large quantity, $N_2O_2$ is evolved.

$$OH_2 + N_2O_3 = 2NOHo$$
$$OH_2 + 3N_2O_3 = 2NOHo + 2N_2O_2$$

**94. Nitrous acid** NOHo produces salts called nitrites. These are frequently found in the well waters of towns, from the oxidation of ammonia. Sodic nitrite was mentioned as a source of nitrogen in Chap. XVI. It is obtained by fusing sodic nitrate in a porcelain crucible. The fused mass decomposes. On cooling and dissolving a portion, it should give an alkaline reaction to test-paper. Nitrous acid sometimes acts as a reducing agent, abstracting oxygen from unstable compounds containing it. At others acting as an oxidising agent.

*Exp.* 17. Add a solution of permanganate of potash to a solution of sodic nitrite, made slightly acid by a few drops of HCl; the colour will be discharged from loss of oxygen.

Sodic nitrite added in the same manner to water, tinted with magenta, acts as an oxidising agent.

**95. Nitric peroxide** $\left\{\begin{matrix} NO_2 \\ NO_2 \end{matrix}\right.$ or $N_2O_4$ has been already prepared by passing nitric oxide into air and oxygen.

*Exp.* 18. Shake up a test-tube of the gas with a solution of hydroxyl; the red fumes disappear and an intensely acid liquid—nitric acid—is produced.

$$\left\{\begin{matrix} NO_2 \\ NO_2 \end{matrix}\right. + Ho_2 = 2NO_2Ho$$

---

## CHAPTER XVIII.

### AMMONIA AND AMMONIC SALTS.

96. AMMONIA is a gas having the composition $NH_3$. There is, however, no known means of making the elements composing it unite directly with each other. It was formerly obtained by the destructive distillation of animal substances which are rich in hydrogen and nitrogen. It is now obtained from the refuse gas liquors which float on the tar, in which it exists as ammonic carbonate. Hydrochloric acid is added to this, ammonic chloride is formed, and carbonic acid escapes. The ammonic chloride being obtained from the solution by evaporation is purified by sublimation. From this ammonic chloride ammonia gas is obtained by heating with slaked lime. The gas is very soluble in water, and must therefore be collected over mercury or by displacement.

$$2NH_4Cl + CaHo_2 = CaCl_2 + 2NH^3 + 2OH_2$$

*Exp.* 1. Rub some ammonic chloride (sal ammoniac) to a powder in a mortar. Then mix with it about the same weight of slaked lime. Put it into a flask with a long delivery-tube bent upwards as in Fig. 41. Add water to make it into a thick paste, and let the delivery-tube from the flask pass to the top of a bottle supported in an inverted position on the ring stand. Heat gently. The gas, which is lighter than air, will rise to the top of the bottle and push the air downwards. Hold a turmeric paper to the mouth of the bottle from time to time, it will be turned brown when the bottle is full. Immerse a taper in the gas, it will be extinguished.

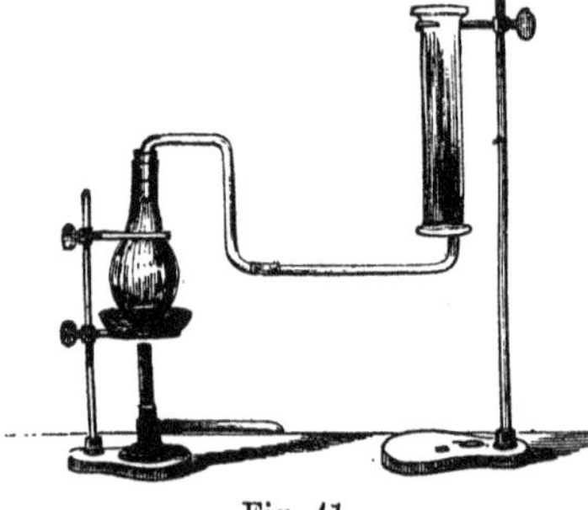
Fig. 41.

Collect another jar of the gas, and a narrow tube of stout glass, as used in *Exp.* 4, Chap. X.

*Exp.* 2. Bring the end of the stout glass tube below the surface of the water. Remove the finger, and shake up the small quantity of water which enters first. Introduce it again into the trough, the water will rush up with great violence. Water dissolves 780 times its volume of the gas.

*Exp.* 3. Prepare a jar of HCl (*see Exp.* 2, Chap. X.), and bring a jar of ammonia over it. Remove the glass plates. The two gases will *combine* together and form a solid white body, ammonic chloride $NH_4Cl$.

A glass rod moistened with HCl held over the bottle of ammonia will produce white fumes of $NH_4Cl$.

*Exp.* 4. Heat the delivery-tube supplying the ammonia gas by the Bunsen flame, the hot gas will inflame at the end of the tube when a light is applied.

The solution of ammonia is obtained by passing the gas through a series of Woulff's bottles provided with safety-tubes. It was formerly obtained by the distillation of stags' horns, hence the common name still retained of spirit of hartshorn. It is sometimes called the volatile alkali. The salts of ammonia are supposed to be built up like those of potassium and sodium, by supposing the existence of a *monad* radical having the composition $NH_4$ as in $NH_4Cl$ given for ammonic chloride in the equation at the commencement of this chapter, and in the chapter on nitrogen. This is similar to KCl and NaCl. The oxide of this body, which has been called **Ammonium,** would be $(NH_4)_2O$, and its hydrate, like potash KHo, would be $NH_4Ho$, which would be the symbol for solution of ammonia or ammonic hydrate.

The abbreviation Am is used for this radical $NH_4$ by Dr Frankland, hence the three bodies named above would be AmCl, $Am_2O$, and AmHo.

*Exp.* 5. Make some sodium amalgam, by gently heating a little mercury in an evaporating dish, and *carefully* putting some bits of the metal sodium into it. The sodium inflames as it amalgamates with the mercury. Make a strong solution of ammonic chloride and pour upon it. The mass swells up, and common salt NaCl is formed in the solution, the hypothetical $NH_4$ amalgamates with the mercury. It speedily decomposes, and nothing but mercury is left.

The salts of ammonia or Ammonium are obtained by adding the acids to the hydrate; it must be done carefully, particularly in the case of sulphuric acid.

$$AmHo + HCl = AmCl + OH_2$$
$$AmHo + NO_2Ho = NO_2Amo + OH_2$$
$$AmHo + SO_2Ho_2 = SO_2Amo_2 + 2OH_2$$

---

## CHAPTER XIX.

### SULPHUR AND SULPHIDES—SULPHURETTED HYDROGEN.

**97. Sulphur** $S^{VI}$ is a hexad element. It is found mixed with earthy matter in Sicily, and also crystallised in cavities. It also exists in combination with metals as sulphides ($FeS_2$ and $Cu_2S$, etc.), and with oxygen and metals as sulphates. The sulphur of commerce is obtained by distilling (sublimation) the *crude* sulphur, the vapours being conveyed into a chamber of brickwork, on the walls of which they condense as "flowers of sulphur." If the walls be hot, the sulphur melts and trickles down into wooden moulds in the floor, forming *roll sulphur* or *brimstone.* When the sulphides of iron and copper (pyrites) are roasted, the sulphur sublimes, and the oxygen of the air converts them into oxides and sulphates.

*Exp.* 1. Heat a bit of galena (lead ore PbS) on charcoal before the blowpipe, the sulphur sublimes and burns with a blue flame. If the bead of lead obtained be heated in the oxidising flame, a yellow incrustation will appear on the charcoal.

*Exp.* 2. Powder some iron pyrites ($FeS_2$) and put it in a hard glass tube closed at one end—or a hard test-tube. Heat it over the Bunsen flame, covering the open end with the thumb. The sulphide will be decomposed, and brownish drops of sulphur will condense on the cool part of the tube.

$$3FeS_2 = Fe_3S_4 + S_2$$

Sulphur is a brittle yellow solid, without taste or smell. It melts at 115°C, forming a *yellow liquid;* at a higher temperature it becomes *brown* and *viscid,* becoming more liquid at 240°C, when it inflames if exposed to the air. At 440° it boils, retaining its brown colour. At a temperature of 1000°C the vapour of sulphur occupies three times the volume it does just above the boiling point; sulphur vapour at 1000°C being thirty-two times the weight of an equal bulk of H. at the same temperature. Hence its atomic weight. Sulphur is hence a *hexad* at the boiling point, but at 1000° a *dyad.*

Sulphur is insoluble in water, but soluble in carbonic disulphide $CS_2$, and also in turpentine. It is also soluble in potassic and calcic hydrates, potassic and calcic sulphide being formed. HCl added to the latter precipitates the sulphur as a white powder—milk of sulphur.

*Exp.* 3. Hold a bit of roll brimstone in the hand, it crackles from the unequal expansion of the interior and exterior. Hold it over the flame of a spirit-lamp, pieces fly off.

*Exp.* 4. Heat some sulphur in a Florence flask, the changes mentioned in a previous paragraph will be seen, vapours being given off which condense in the neck of the flask. Whilst in the *viscid* condition, pour it out of the flask into water round an inverted funnel. Translucent strings, in a *plastic* condition, may be taken out of the water and moulded like wax.

*Exp.* 5. Melt about a quarter of a pound of sulphur in a Cornish crucible on a clear fire. Cover it with a bit of tile before it has all melted, or it will inflame, combining with the oxygen of the air to form $SO_2$. When it has all melted, take it off the fire with a pair of crucible tongs; let it cool, and when a crust has formed over the surface, make a hole in it and pour out the liquid sulphur.

Beautiful **prismatic crystals** will form in the interior. When the crucible is quite cold, break it with a hammer and they will be seen better

*Exp.* 6. Dissolve some flowers of sulphur in carbonic disulphide ($CS_2$); or turpentine, if the latter unpleasant smelling liquid is not at hand; in either case keeping away from a light. Pour the solution into an evaporating dish to evaporate slowly. Crystals of a totally different shape—**octahedrons**—like double pyramids, will separate.

As sulphur crystallises in two different forms, it is said to be *dimorphous.* The crystals obtained from fusion, however, become opaque, and break up into *octahedrons* in a few days. The *plastic* variety also returns to the ordinary yellow kind, but while in its plastic brown state it is insoluble in $CS_2$. In consequence of these differences, sulphur is said to exist in three allotropic states.

Sulphur combines with the metals, forming sulphides, which are similar in their composition to the oxides. The metals in the form of leaf or fine wire burn in sulphur vapour.

*Exp.* 7. Melt a little more sulphur in the flask used in *Exp.* 4, or in a large test-tube. Keep the vapours given off *hot,* by letting a flame impinge on the upper part of the flask or tube. Attach a bit of Dutch metal to a wire, and plunge it in the heated vapour; it will take fire, cupric sulphide $CuS_2$ being formed. Other metals in the state of leaf or foil may be substituted, or a bundle of fine iron or copper wire.

*Exp.* 8. Shoot some iron filings into the flask of melted sulphur, ferrous sulphide FeS will be formed. This compound is used in the making of sulphuretted hydrogen. It is made by putting lumps of iron into a crucible containing melted sulphur, an atom of Fe only combining with one atom of sulphur; both act as *dyads.*

### 98. Sulphuretted hydrogen $SH_2$ (H – S – H.)

Sulphur can be made to unite directly with hydrogen, but the compound $SH_2$ is usually obtained by the action of HCl or $SO_2Ho_2$ upon *ferrous* sulphide FeS just mentioned, or by the action of HCl on antimonious sulphide $Sb_2S_3$ at a gentle heat.

$$FeS + 2HCl = FeCl_2 + SH_2$$
$$FeS + SO_2Ho_2 = SO_2Feo'' + SH_2$$
$$Sb_2S_3 + 6HCl = 2SbCl_3 + 3SH_2$$

*Exp.* 9. Break some ferrous sulphide into small pieces, and put it into the hydrogen apparatus; or, if a stink-closet be at hand, keep a bottle for the purpose. Cover it with water, and pour in some HCl; a gas having an offensive odour like rotten eggs comes off. It is rather soluble in cold water, and should therefore be collected in jars over warm water.

*Exp.* 10. Leave a little water in one of the jars, and shake it up with the gas; the hand will be sucked in, showing that the water has dissolved.

Put a blue litmus paper in the solution, it will be reddened, showing the gas to have acid properties. It is sometimes called hydrosulphuric acid. Keep the solution for future experiments.

*Exp.* 11. Inflame a jar of the gas, and hold a dry glass over it; water and sulphurous anhydride will be produced.

Or, burn a jet of the gas previously dried by passing it through a tube containing chloride of calcium. (*See Exp.* 9, Chap. VIII.)

*Exp.* 12. Graduate the explosion tube into five parts. Fill it with two parts sulphuretted hydrogen * and three of oxygen. Bring the mouth of the tube to a flame, the mixed gases will explode violently, water and sulphurous anhydride being produced.

$$2SH_2 + 3O_2 = 2SO_2 + 2OH_2$$

*Exp.* 13. Prepare a cylinder of chlorine, and bring it over an equal-sized cylinder of $SH_2$. Remove the glass plates, and let the gases mix. The $SH_2$ will be decomposed by Cl, which combines with the hydrogen to form HCl, sulphur being deposited round the sides of the jar. (The latter is easily cleansed with potash.) The experiment may be varied by adding chlorine water to a solution of $SH_2$†.

*Exp.* 14. Prepare a jar of sulphurous anhydride (*see* next chapter, *Exp.* 1), and bring it over a similar sized jar of $SH_2$. The gases will decompose each other, after the following equation—

$$4SH_2 + 2SO_2 = 4OH_2 + 3S_2$$

*Exp.* 15. Pass $SH_2$ from the delivery-tube, to which a bit of flexible tube is attached through *solutions* of the following salts—

* The smell of sulphuretted hydrogen is readily destroyed by burning sulphur.

† The generation of a *little* chlorine speedily destroys the odour of $SH_2$ in a room.

using a fresh bit of glass tube at the extremity, in each case, and note the results, viz. :—

*Sulphate of copper, plumbic acetate, arsenious anhydride, tartar emetic (tartrate of antimony), sulphates of zinc and chromium and ferric chloride.*

The solution of the gas will do instead of the gas itself. The solution, however, on exposure to the air, loses its smell, water being formed by the oxidation of the hydrogen, and sulphur being deposited.

Sulphuretted hydrogen is used in *Analytical Chemistry*, to separate metals into groups by precipitation, mostly as sulphides. The sulphides which are insoluble in water may be precipitated from their ordinary aqueous solutions; those insoluble in acids form acid solutions also; but those which are soluble in acids must first be made alkaline by the addition of ammonia. The sulphides of K, Na, Ba, Sr, Ca, and Mg, being soluble, cannot be precipitated by sulphuretted hydrogen.

---

## CHAPTER XX.

### Compounds of Sulphur with Oxygen and Hydroxyl.

**99.** There are two compounds of sulphur with oxygen, sulphur*ous* anhydride $O = S = O$ or $SO_2$, and sulphur*ic* anhydride $O = S = O$ or $SO_3$.

O

These bodies added to water form two very important compounds, sulphur*ous* and sulphur*ic* acids.

**Sulphurous anhydride** $SO_2$ has been already prepared by burning sulphur in oxygen, and as a secondary product in many instances. It is a gaseous body. Its weight compared with H. and air would be found thus:

$$S + O_2$$

$$\frac{32 + 32}{2} = \frac{64}{2} = 32, \text{ its volume weight.}$$

And $32 \div 14{\cdot}47 = 2{\cdot}21$ times the weight of air.

It is usually prepared by heating copper or mercury with sulphuric acid, the acid being deoxidised, gaseous $SO_2$ being evolved, water and sulphates of the metals being also formed.

$$2SO_2Ho_2 + Cu = SO_2Cuo'' + 2OH_2 + SO_2$$

*Exp.* 1. Into a flask put some copper clippings and concentrated sulphuric acid. Heat gently over a sand-bath, and a gas with the odour of burning sulphur will be given off. It is very soluble in water, and must therefore be collected over mercury or by downward displacement. A blue litmus will be reddened at the opening where the delivery-tube enters, as the gas has *acid* properties. Collect several jars of the gas.

*Exp.* 2. Immerse a taper in a jar of the gas, it will be extinguished, and the gas does not inflame.

*Exp.* 3. If a jar of the gas has been collected over mercury, let it stand upright in the bath supported by a clip of the retort stand. Pour some water, coloured with litmus, on the mercury, and then raise the jar out of the mercury into the water. The gas will dissolve very rapidly. Water dissolves about 40 times its volume of the gas.

Fig. 42.

A solution of the gas may be made by passing the gas into cold water till no more is absorbed.

*Exp.* 4. Put a moistened bunch of violets into a jar of $SO_2$, they will be bleached, but the colour will be restored if they be dipped into a weak solution of acid. Straw plait, wool, and silk are bleached by exposing them to the vapours of burning sulphur, as chlorine would injure such fabrics.

Infusion of rose leaves may be used, and the colour be restored by dilute sulphuric or hydrochloric acids.

*Exp.* 5. Pass a stream of $SO_2$ into a test-tube, surrounded by a freezing mixture,—crushed ice, or snow, and salt. The gas will be condensed into a colourless liquid.

The solution of the gas in *Exp.* 3, produces sulphur*ous* acid, $SOHo_2$, and this, upon being cooled to the freezing point, deposits white crystals. The gas when passed into solutions of metallic hydrates forms two classes of *salts*, called *sulphites*. Sulphurous acid has the same effect.

$$NaHo + SO_2 = SOHoKo, \text{ or } NaHo + SOHo_2 = SOHoKo + OH_2$$
$$\text{and } 2NaHo + SO_2 = SOKo_2, \text{ or } 2NaHo + SOHo_2 = SONao_2 + 2HO_2$$

**100. Sulphuric acid.** This most important acid may be prepared in the following ways. Its preparation on a large scale in leaden chambers may be passed over in an elementary treatise.

*Exp.* 6. Expose some of the solution of $SO_2$ *i.e.* $SOHo_2$ to the air; it will absorb oxygen, and be converted into sulphuric acid. This may be tested by adding a solution of baric nitrate to a freshly made solution of $SO_2$; very little white pp. will be produced. Added to that which has been exposed to the air for several hours, a dense white pp. of baric sulphate will be obtained.

*Exp.* 7. Boil some flowers of sulphur with nitric acid. The sulphur will be oxidised at the expense of the nitric acid, and sulphuric acid will be obtained. When cold, the liquid will not dissolve copper, but when heated will be again deoxidised with evolution of $SO_2$.

*Exp.* 8. Heat liquid $SOHo_2$ with nitric acid, the same result.

Sulphuric acid may be considered as made up of a body called sulphuric anhydride $SO_3$ and water, or of sulphurous anhydride $SO_2$ and hydroxyl $Ho_2$.

$$SO_3+OH_2=SO_2Ho_2, \text{ or } SO_2+Ho_2=SO_2Ho_2$$

Sulphuric anhydride may be obtained by heating Nordhausen sulphuric acid, and condensing the vapours in a flask surrounded by a freezing mixture; or, by passing $SO_2$ and oxygen over ignited spongy polatinum as in the following:—

*Exp.* 9. Fit up a three-necked Woulff's bottle, so that a stream of $SO_2$ shall pass in by one tube, oxygen by a second, and let the third neck be connected with a bit of combustion tube containing a small quantity of *spongy platinum.* The gases will only mix in the Woulff's bottle, but when the platinum is heated, the mixed gases which condense in it enter into combination, and vapours of sulphuric anhydride will pass out. If sufficient quantity be obtained in white crystalline needles, they may be allowed to pass into a test-tube containing water, which they will do with a hissing noise, forming sulphuric acid.

*Exp.* 10. Shake up a mixture of sulphurous acid (solution of $SO_2$) with a solution of hydroxyl mentioned on page 100. Sulphuric acid will be obtained. Test as in *Exp.* 6.

*Exp.* 11. Into a three-necked Woulff's bottle containing hot water, pass a stream of nitric oxide and oxygen gas, and as the red vapours appear, a stream of $SO_2$. The $N_2O_3$ will be converted into $N_2O_4$, which will be again deoxidised by the $SO_2$, and the resulting compound will be dissolved by the stream, and form a solution of sulphuric acid.

Sulphuric acid forms salts called sulphates, by the displacement of the whole or part of the hydroxyl, by a *metallic radical.* They are also formed by the action of the acid on metallic carbonates. The acid chars organic substances—abstracting the elements of water, and leaving the carbon.

*Exp.* 12. Make a strong solution of loaf-sugar in a beaker, and pour concentrated sulphuric acid into it. A dense black mass will froth up, and most likely overflow the vessel, which should be placed in a dish.

If sulphur be boiled with a solution of sodic sulphite, a salt called sodic hyposulphite $SSONao_2$ is formed. This is supposed to contain hyposulphurous acid $SSOHo_2$, which decomposes on the addition of dilute sulphuric acid.

*Exp.* 13. Make a solution of sodic hyposulphite (used by photographers for dissolving the excess of silver salt), and add dilute sulphuric acid; $SO_2$ will be evolved, which may be detected by its odour, and yellow sulphur will be deposited.

*Exp.* 14. Add dilute sulphuric or hydrochloric acid to sodic sulphite, $SO_2$ will be evolved; but no sulphur will be deposited.

---

The student will now be able to understand the various "Modes of chemical action." * A few examples will be given, and other similar cases should be sought out.

Chemical change takes place—

1. When elements or compounds combine *directly* with each other.

$$S + O_2 = SO_2$$
$$N_2O_2 + O_2 = N_2O_4$$
$$NH_3 + HCl = NH_4Cl$$

2. When compounds are split up into their elements, or into less complex compounds.

$$OAg_2 = Ag_2 + O$$
$$NONH_4O = N_2 + 2OH_2$$
$$NO_2NH_4O = ON_2 + 2OH_2$$

* *See* "Frankland's Lecture Notes," and Valentin's "Practical Chemistry."

3. When one element or group of elements displaces another element or group of elements.

$$2HCl + K_2 = 2KCl + H_2$$
$$SO_2Cuo'' + Fe = SO_2Feo'' + Cu$$
$$Pb\bar{A} + Zn = Zn\bar{A} + Pb$$

4. When elements or groups of elements in one body are exchanged for other elements or groups of elements in another body.

$$CuCl_2 + SH_2 = CuS + 2HCl$$
$$CuCl_2 + 2AmHo = CuHo_2 + 2AmCl$$
$$\begin{matrix}NO_2\\NO_2\end{matrix}Bao'' + SO_2Ho_2 = SO_2Bao'' + 2NO_2Ho$$
$$COCao'' + 2HCl = CaCl_2 + OH_2 + CO_2$$

5. When the elements of a compound are re-arranged, as in the conversion of starch into sugar. Numerous instances occur in ORGANIC CHEMISTRY.

# QUESTIONS

## REQUIRING A PRACTICAL KNOWLEDGE OF CHEMISTRY.

1. The following metals are heated in the Bunsen flame :—*Magnesium, platinum, zinc, sodium, copper;* what is the effect upon each, and what products would be formed ?

2. When *sodium* is thrown into water, what takes place ?

3. Describe an experiment to show the difference between a *mechanical mixture* and a *chemical compound.*

4. What is the effect of heat upon *manganic oxide, potassic chlorate, ammonic chloride, ammonic nitrate, phosphorus,* and *sulphur* respectively ?

5. What is the action of HCl on the metals Zn, Fe, and Mg ? Express the reactions in equations, and draw graphic formulæ of the products formed.

6. You have given to you zinc, sulphuric acid, caustic, potash, and water, and are required to prepare hydrogen from these materials by two distinct processes ; state how you would proceed, and show by an equation the chemical changes in each case. (Exam. 1869).

7. I have two cylindrical jars of hydrogen ; I hold one of them mouth upwards, and the other mouth downwards. At the expiration of half a minute I plunge a lighted taper into each jar ; describe exactly what you would expect to take place in each case. (Exam. 1872).

8. You have given to you some potassic chlorate (chlorate of potash) and black oxide of manganese, and are required to make and collect oxygen from these materials. Describe minutely how you would do it, and make a sketch of the apparatus which you would employ. What are the chief properties of oxygen ?

(Question set, 1872, for those who wished to show a good knowledge of laboratory practice.)

9. What compounds can be formed from the following elements :—Oxygen, hydrogen, chlorine ? and how ?

10. What elements can be obtained from hydrochloric acid, ammonia, and water ? Say how you would obtain them in each case.

11. When hydrogen is burnt in chlorine, what is produced? Suppose it were burnt in air what product is formed ?

12. I mix together chlorine and hydrogen, and expose the mixture to sunlight ; what happens ? At the termination of the

experiment I add ammonia to the product. What is the name and formula of the compound formed? (Exam. 1871).

13. If you had given to you some common salt and sulphuric acid, and were required to fill a glass jar with HCl, describe how you would do it, and give a sketch of the apparatus you would employ. (Exam. 1872).

14. What is the action of sulphuric acid upon *copper, sodic chloride*, and *manganic oxide* respectively when assisted by heat? If the two latter are acted upon together what then takes place?

15. If a mixture of manganic oxide (peroxide of manganese) and hydrochloric acid be heated, what chemical change takes place? Give the name and properties of the gas which is evolved. (Exam. 1869).

16. An electric spark is passed through a mixture containing 120cc of hydrogen and 60cc of oxygen. How would you arrange the experiment so as to show the gaseous condensation?

17. What is the action of the electric current upon water, solutions of HCl and $NH_3$? State minutely how you would perform the experiments to show the results.

18. How would you prepare a solution of hydrochloric acid?

19. How would you prove the solubility of hydrochloric acid, ammonia, and carbonic anhydride gases?

20. If I agitate with pure water a mixture of hydrogen, nitrogen, carbonic anhydride, and hydrochloric acid gases, what effect will be produced? (Exam. 1872).

21. Steam and chlorine are passed through a porcelain tube, heated to redness in a furnace, what takes place? Give equations to show the reaction.

22. You are required to make oxygen from chlorine and water. Describe exactly how you will do it, and give a sketch of the apparatus which you propose to employ. (Exam. 1871).

23. How is potassic chlorate usually prepared? Explain fully the various steps in the process.

24. A piece of borax, a bit of chalk, and some nitre, are respectively heated upon charcoal before the blowpipe flame; what takes place in each case?

25. You have some mercury, a glass flask, and a piece of hard glass tube, and are required to make pure oxygen gas, how will you do it? (Exam. 1869).

26. A stream of $CO_2$ is passed into the following:—*Distilled water, lime water, ammonia solution*, and *baric peroxide* suspended in water; what takes place in each case?

27. How would you prepare boric acid from borax? What is the action of boric acid ($BHo_3$) on sodic carbonate?

28. When nitric acid is poured upon copper, how does the action differ from a simple solution?

29. You have some ammonic carbonate (carbonate of ammonia) and nitric acid, and are required to make and collect laughing gas from these materials; how would you do it? Describe minutely the apparatus you would employ, and make a sketch of it. What are the chief properties of laughing gas?

(Question set, 1871, to show good knowledge of laboratory practice.)

30. You have given to you some sulphur, water, and nitric acid; describe how you would make sulphuric acid from these materials. (Exam. 1871).

31. State exactly how you would prepare nitrogen, nitric oxide, and ammonia. Give equations. (Exam. 1869).

32. How is carbonic oxide usually prepared? Give a sketch of the apparatus you would employ, and the reactions in the process.

33. When carbonic oxide burns in air, what product is formed? How would you test the properties of the gas and product?

34. When sulphuretted hydrogen is burnt in air, what products are formed? How would you prove their formation?

35. What is the action of dilute sulphuric acid upon sodic sulphite, and also upon sodic hyposulphite?

36. Describe some experiments to show "the modes of chemical action."

# APPARATUS AND MATERIALS

REQUIRED TO PERFORM THE EXPERIMENTS IN THE BOOK.

## A.—List of Apparatus Recommended by the Science and Art Department for each Student.

| | *s.* | *d.* |
|---|---|---|
| Conical Brass Blowpipe, 2s, or Black Japanned, - - | 0 | 8 |
| 6 inches Platinum Wire, - - - - - - | 0 | 6 |
| Platinum Foil, 2 in. long, 1 in. wide, - - - - | 1 | 0 |
| Test-tube Stand, 24 holes, - - - - - - | 2 | 0 |
| 18 Test Tubes, 6 in. × $\frac{3}{4}$ in., - - - - - - | 1 | 6 |
| 12 ,, 5 × $\frac{1}{2}$, - - - - - - - | 0 | 8 |
| 2 Boiling Tubes (large Test Tubes, 8 × $1\frac{1}{4}$), - - - | 0 | 4 |
| 2 Test Tube Brushes, - - - - - - - | 0 | 5 |
| Set of 3 Beakers, - - - - - - - - | 1 | 0 |
| German Flasks, 1 of each, 2 oz., 4 oz., 8 oz., 16 oz., 30 oz., | 1 | 9 |
| Berlin Porcelain Crucible, $1\frac{1}{2}$ in. diameter, - - - | 0 | 4 |
| Evaporating Dishes, $2\frac{3}{4}$ in. and $3\frac{1}{2}$ in. diameter, - - | 0 | 11 |
| Funnels, one of each, $1\frac{1}{2}$ in., 2 in., and 3 in. diameter, - | 0 | 10 |
| English Filter Papers, 2 pkts. of 100, $2\frac{3}{4}$ in. and $4\frac{1}{2}$ in. dia. | 0 | 10 |
| Iron Retort Stand, 7s., or smaller one, - - - - | 3 | 6 |
| Wire Gauze, 2 pieces, 5 in. sq., - - - - - | 0 | 4 |
| Tin Plate Sand Bath, 5 in. diameter, - - - - | 0 | 4 |
| 6 Watch Glasses, - - - - - - - - | 0 | 6 |
| $\frac{1}{2}$ lb. Soft Glass Tube, $\frac{3}{16}$ to $\frac{1}{4}$ in. diameter, - - - | 0 | 8 |
| $\frac{1}{2}$ lb. Combustion Tube, $\frac{3}{8}''$, - - - - - - | 0 | 6 |
| $\frac{1}{4}$ lb. Glass Rod, $\frac{3}{16}''$ bore, - - - - - - | 0 | 3 |
| 2 feet Black Caoutchouc Tube, $\frac{3}{8}''$ bore, - - - - | 1 | 4 |
| 2 feet ,, ,, $\frac{1}{8}''$ ,, - - - - | 0 | 8 |
| Thistle Funnel, - - - - - - - - | 0 | 3 |
| 3 dozen Assorted Corks, - - - - - - - | 1 | 4 |
| Woulfe's Bottle, 2 necks, $\frac{1}{2}$ pint size, - - - - | 1 | 0 |
| Stoppered German Retort, 2 oz., - - - - - | 0 | 8 |
| Set of 3 Cork Borers, $\frac{3}{16}$, $\frac{1}{4}$, and $\frac{3}{8}$, with iron rod, - - | 1 | 0 |
| Triangular File and Handle, - - - - - - | 0 | 8 |
| 5-inch Round File and Handle, - - - - - | 0 | 9 |
| Bunsen's Gas Burner, with Blowpipe Jet, Star Support, Chimney, and Rose Burner, - - - - - - | 3 | 6 |
| Iron Crucible Tongs, - - - - - - - | 1 | 0 |
| 4-inch Porcelain Mortar and Pestle, - - - - | 1 | 0 |
| Books of Test Papers, Blue and Red Litmus, etc., - - | 0 | 6 |
| Solution of Cobaltous Nitrate, $\frac{1}{2}$ oz. in Stoppered Bottle, | 0 | 7 |
| ,, Argentic ,, $\frac{1}{2}$ oz. ,, - | 0 | 6 |
| ,, Platinic Chloride, $\frac{1}{2}$ oz. ,, - | 1 | 9 |
| 1 Pint Methylated Alcohol in Bottle, - - - - | 1 | 6 |
| Glass Spirit Lamp, 4 oz. capacity, - - - - - | 1 | 0 |
| Deal Box, to contain the Set of Apparatus, - - - | 2 | 0 |

## B.—Other Chemical Apparatus Required for Use of a Class.

| | s. | d. |
|---|---|---|
| 3 Cylindrical Gas Tubes, 6″ × 1¼″, - - - - - - | 2 | 0 |
| 1 Deflagrating Jar, - - - - - - - - - - | 1 | 0 |
| 1 ,, stoppered 10 oz., - - - - - | 1 | 3 |
| Cap and Spoon (extra spoon 6d.), - - - - - - | 1 | 0 |
| Explosion Tube (or, Soda Water Bottle), - - - | 1 | 6 |
| 2 Chloride of Calcium Tubes, - - - - - - - | 1 | 0 |
| Balloon of Goldbeater's Skin, - - - - - - - | 1 | 0 |
| Clark's Retort, - - - - - - - - - - | 0 | 6 |
| Glass Dishes (basins can be used by students) each 10d. to | 2 | 0 |
| Pneumatic Trough (Japanned Tin), 1 gallon 3s.; 2 gallon | 5 | 0 |
| * Mercury Trough, - - - - - - - - - - | 1 | 9 |
| * 4 lbs. Mercury for Trough, - - per lb. 2s. 6d. to | 4 | 6 |
| ** Eudiometer, straight, graduated in c. c. - - 6s. to | 9 | 0 |
| A Wooden Vice, - - - - - - - - - - | 2 | 6 |
| A 3-necked Woulfe's Bottle, pint size, - - - - - | 1 | 6 |
| A Nest of 3 Cornish Crucibles, - - - - - - | 0 | 3 |
| A Box of Scales and Weights, - - - - - - | 2 | 6 |
| Maw's Set of Gramme Weights, - - - - - - | 1 | 0 |
| Herepath's Blowpipe, to use with Bunsen's Burner, - | 2 | 6 |
| Half Round Rasp in Handle, - - - - - - - | 1 | 0 |

## C.—*Electrical Apparatus Necessary to Perform some of the Experiments.

| | s. | d. |
|---|---|---|
| Pair of 6-inch Bar Magnets, - - - - - - - - | 2 | 6 |
| 5, 4-inch Cells Bunsen's Battery, complete, - - - | 22 | 0 |
| 2 V-Tubes on Stands, with Platinum Plates and Wires (page 49), - - - - - - - each | 6 | 6 |
| Voltameter for Mixed Gases (fig. 17, page 50), - - | 5 | 6 |
| ,, with 2 Tubes, for collecting Gases separately (fig. 18, page 51), - - - - - - - | 9 | 6 |
| ** Induction Coil, - - - - - - - - - - | 10 | 0 |
| ** Or, Electrophorus, - - - - - - - - - | 5 | 6 |

## D.—Materials Required to Perform the Experiments for a Class of Students.

| | s. | d. |
|---|---|---|
| 1 lb. Granulated Zinc, - - - - - - - - - | 0 | 8 |
| ½ lb. Oxygen Mixture, - - - - - - - - - | 1 | 0 |
| 1 lb. Black Oxide Manganese, - - - - - - - | 0 | 3 |
| ¼ lb. Chlorate Potash, - - - - - - - - - | 0 | 8 |
| ½ oz. Mercuric Oxide, - - - - - - - - - | 0 | 3 |
| ¼ lb. Copper Turnings, - - - - - - - - - | 0 | 6 |
| ¼ lb. Ferrous Sulphide (Sulphide of Iron), - - - - | 0 | 2 |
| ½ lb. Chloride Calcium (for drying gases, in stoppered bottle), | 0 | 3 |

| | s. | d. |
|---|---|---|
| ½ lb. Sulphur, | 0 | 2 |
| 1 oz. Phosphorus, | 0 | 4 |
| 1 oz. Metal Sodium, | 1 | 0 |
| ½ dr. ,, Potassium, | 1 | 0 |
| ¼ oz. Iodine, | 1 | 2 |
| ½ lb. Ammonia Alum, | 0 | 2 |
| 2 oz. Chloride Ammonium, | 0 | 4 |
| ¼ lb. Borax, | 0 | 5 |
| ½ oz. Boracic Acid, | 0 | 2 |
| 2 oz. Red Lead, | 0 | 1 |
| 1 oz. Acetate Lead, | 0 | 1 |
| 1 oz. Sulphate Zinc, | 1 | 0 |
| ½ lb. Sulphate Soda (Glauber's Salts), | 0 | 2 |
| ¼ lb. Sulphate Copper, | 0 | 2 |
| ¼ lb. Sulphate Iron, | 0 | 2 |
| ¼ lb. Nitre (Saltpetre, Nitrate Potash), | 0 | 2 |
| ¼ lb. Nitrate Soda (Cubic Nitre), | 0 | 2 |
| 2 oz. Baric Chloride or Nitrate, | 0 | 2 |
| 1 oz. Cupric Chloride, | 0 | 3 |
| 1 oz. Cupric Oxide, | 0 | 3 |
| 1 oz. Ferric Chloride, | 0 | 6 |
| ¼ lb. Caustic Soda, | 0 | 2 |
| 1 oz. Caustic Potash, | 0 | 3 |
| ¼ oz. Peroxide Barium, | 0 | 3 |
| ½ oz. Sodic Nitrite, | 0 | 4 |
| 1 oz. Ammonic Nitrite, | 0 | 3 |
| 2 oz. Ammonic Nitrate, | 0 | 5 |
| 1 oz. Oxalic Acid, | 0 | 2 |
| ¼ oz. Pyrogallic Acid, | 1 | 0 |
| ½ oz. Ferrocyanide Potassium, | 0 | 3 |
| ¼ oz. Mercuric Chloride, | 0 | 2 |
| ¼ oz. Iodide Potassium, | 1 | 0 |
| 1 oz. Metallic Antimony, | 0 | 2 |
| Leaf Gold and Dutch Metal, Book of each | 2 | 0 |
| 1 lb. Common Hydrochloric Acid, | 0 | 2 |
| 1 lb. ,, Sulphuric ,, | 0 | 2 |
| ½ lb. ,, Nitric ,, | 0 | 4 |
| ½ lb. Ammonia Solution ,, | 0 | 6 |

The last four, in Winchester pint stoppered bottles at 6d. each; the other materials in 4 oz. Flint Glass wide-mouthed stoppered bottles at 5s. 6d. per dozen, or common wide-mouthed bottles for corking, at 1s. or 2s. per dozen.

*** The Chemical Apparatus and Materials can be had of M. Jackson, 65 Barbican, London, E.C.; G. Mason, 39 Union Street, Glasgow; or from the Publishers. The Electrical Apparatus of W. Peters, Philosophical Instrument Maker, 36 Whiskin Street, Clerkenwell, E.C., at the prices named.

www.ingramcontent.com/pod-product-compliance
Lightning Source LLC
LaVergne TN
LVHW021414110826
845150LV00007B/1914